AF352857

Mechanical Properties of Materials

Mechanical Properties of Materials

Editors

Giovanni Bruno
Alberto De Stefani
Antonio Gracco

MDPI • Basel • Beijing • Wuhan • Barcelona • Belgrade • Manchester • Tokyo • Cluj • Tianjin

Editors

Giovanni Bruno
University of Padua
Italy

Alberto De Stefani
University of Padua
Italy

Antonio Gracco
University of Padua
Italy

Editorial Office
MDPI
St. Alban-Anlage 66
4052 Basel, Switzerland

This is a reprint of articles from the Special Issue published online in the open access journal *Applied Sciences* (ISSN 2076-3417) (available at: https://www.mdpi.com/journal/applsci/special_issues/mechanical_properties_materials).

For citation purposes, cite each article independently as indicated on the article page online and as indicated below:

LastName, A.A.; LastName, B.B.; LastName, C.C. Article Title. *Journal Name* **Year**, *Volume Number*, Page Range.

ISBN 978-3-0365-1084-2 (Hbk)
ISBN 978-3-0365-1085-9 (PDF)

Contents

About the Editors

Giovanni Bruno studied Dentistry at the University of Padua in 2017 with the maximum evaluation cum laude, discussing original research on dental agenesis in patients affected by cleft lip and palate. In 2016, he was a visiting student at the Craniofacial Center at the University of North Carolina. He has participated in several post-graduate courses in orthodontics and pediatric dentistry and has published more than 40 research articles in international journals. He is currently enrolled in a three-year Master of Science program in pediatric dentistry that ends in 2021.

Alberto De Stefani studied Dentistry at the University of Padua in 2017 with the maximum evaluation cum laude, discussing original research on skeletal age evaluation in growing patients. In 2016, he was a visiting student at the Craniofacial Center at the University of North Carolina. He has participated in several post-graduate courses in orthodontics and pediatric dentistry and has published more than 40 research articles in international journals. He is currently enrolled in a three-year Master of Science program in pediatric dentistry that ends in 2021.

Antonio Gracco graduated from the University of Padua in 2003 and specialized in orthodontics in 2006 at the University of Ferrara with the maximum evaluation. He coordinated research activities in orthodontics at the University of Ferrara and, later, at the University of Padua. He is currently associate professor at the University of Padua and Head of the Pediatric Dentistry Master of Science program. He has published more than 100 original pieces of research in international journals.

Preface to "Mechanical Properties of Materials"

In the oral environment, restorative and prosthetic materials and appliances are exposed to chemical, thermal and mechanical challenges. The mechanical properties of a material define how it responds to the application of a physical force. The mechanical properties which are of importance in dentistry include brittleness, compressive strength, ductility, elastic modulus, fatigue limit, flexural modulus, flexural strength, fracture toughness, hardness, impact strength, malleability, Poisson's ratio, shear modulus, shear and tensile strength, torsional strength and Young's modulus. All of these are measures of the resistance of materials to deformation, crack or fracture under an applied force or pressure. Measured responses can be both elastic (reversible on force removal) and plastic (irreversible on force removal). Recent advances in nanotechnology and 3D printing have rapidly spread and manufacturers continuously develop new materials and solutions to provide high-quality dental care, with particular attention being paid to long-term follow-up. Restorative dentistry, prosthodontics, oral surgery, implants, periodontology and orthodontics are all involved in this continuing evolution. This Special Issue focuses on all the recent technology that can enhance the mechanical properties of materials used in all of the different branches of dentistry.

Giovanni Bruno, Alberto De Stefani, Antonio Gracco
Editors

Article

Comparison of the Effects Caused by Three Different Mandibular Advancement Devices on the Periodontal Ligaments and Teeth for the Treatment of Osa: A Finite Element Model Study

Giovanni Bruno [1,*], Alberto de Stefani [1,*], Manila Caragiuli [2], Francesca Zalunardo [1], Alida Mazzoli [3], Daniele Landi [2], Marco Mandolini [2] and Antonio Gracco [1]

[1] Department of Neuroscience, Section of Dentistry, University of Padua, 35100 Padua, Italy; zalunardo.francesca@gmail.com (F.Z.); antonio.gracco@unipd.it (A.G.)

[2] Department of Industrial Engineering and Mathematical Sciences, Università Politecnica delle Marche, 60131 Ancona, Italy; m.caragiuli@staff.univpm.it (M.C.); d.landi@staff.univpm.it (D.L.); m.mandolini@staff.univpm.it (M.M.)

[3] Department of Materials, Environmental Sciences and Urban Planning, Università Politecnica delle Marche, via Brecce Bianche 12, 60131 Ancona, Italy; a.mazzoli@staff.univpm.it

* Correspondence: giobruno93@gmail.com (G.B); alberto.de.stefani@hotmail.it (A.d.S.)

Received: 29 July 2020; Accepted: 27 September 2020; Published: 3 October 2020

Abstract: AIM: The purpose of this study is to compare the stress effects developed on the periodontal ligaments and teeth by three different types of mandibular advancement devices (MADs) using a finite element method (FEM) analysis. Introduction: Obstructive sleep apnea (OSA) is a disease with a high prevalence and, in recent years, the use of MADs as an alternative or support treatment to the continuous positive airway pressure (CPAP) has spread. Their use finds relative contraindications in the case of partial edentulism and severe periodontal disease. Given the widespread of periodontal problems, it is essential to know the effects that these devices cause on the periodontal ligament of the teeth. Materials and methods: Starting from the computed tomography (CT) scan of a patient's skull, 3D reconstructions of the maxilla and mandible were implemented. Three different MADs were prepared for the patient, then 3D scanned, and lastly, coupled with the 3D models of the jaws. The devices have two different mechanics: One has a front reverse connecting rod (Orthoapnea[TM]), and two have lateral propulsion (Somnodent[TM] and Herbst[TM]). A FEM analysis was performed to calculate the stress applied on periodontal ligaments, on every single tooth and the displacement vectors that are generated by applying an advancement force on the mandible. Results: Herbst[TM] and Somnodent[TM] devices present very similar stress values, mainly concentrated on lateral teeth, but in general, the forces are very mild and distributed. The maximum stresses values are 3.27 kPa on periodontal ligaments and 287 kPa on teeth for Somnodent[TM] and 3.56 kPa on periodontal ligaments and 302 kPa on teeth for Herbst[TM]. Orthoapnea[TM] has, instead, higher and concentrated stress values, especially in the anterior maxillary and mandibular area with 4.26 kPa and 600 kPa as maximum stress values, respectively, on periodontal ligaments and teeth. Conclusions: From the results, it is concluded that devices with a bilateral mechanism generate less and more distributed stress than an anterior connecting rod mechanism. Therefore, they may be advisable to patients with compromised periodontal conditions in the anterior area.

Keywords: dental materials; orthodontics; obstructive sleep apnea; mandibular advancement device; finite element method

---

1. Introduction

Obstructive sleep apnea (OSA) is characterized by various and recurring episodes of reduction (hypopnea) or cessation (apnea) of the airflow during sleep. It results from the obstruction due to the collapse of the upper airway [1]. The pathophysiology of OSA is multifactorial and includes a reduction in upper airway dimensions that is caused by both anatomical and functional alterations (obesity or maxillofacial structural changes) and increased pharyngeal collapsibility due to reduced neuromuscular compensation and lack of the pharyngeal protective reflex during sleep [2–6].

OSA is one of the most prevalent chronic respiratory disorders. In recent population-based studies, the estimated prevalence of moderate to severe sleep-disordered breathing ranges from 3% to nearly 50% depending on the age group and sex. Recent studies suggest an increase in the incidence that is probably related to the growing prevalence of overweight and obese individuals [7,8].

Extensive population-based studies have shown that untreated moderate or severe OSA is associated with severe complications. The most important are the increased cardiovascular and cerebrovascular disease, stroke, chronic inflammation, hypertension, and metabolic syndrome resulting in increased morbidity and mortality. The main symptoms are snoring, day sleepiness, nocturia, frequent nocturnal waking caused by choking or gasping sensations, psychological or cognitive dysfunction, morning headaches, low concentration, irritability, and erectile dysfunction. Many patients report a reduction in their life quality due to obstructive sleep apnea [9–13].

OSA is an under-estimated problem even though there are many screening tools such as the STOP-BANG questionnaire, Epworth scale, and Berlin questionnaire [14–17]. The gold standard for the diagnosis of OSA is polysomnography (level I study), which consists of an instrumental examination carried out in dedicated structures.

Polysomnography involves the collection of seven or more data channels, including respiratory, electrocardiogram, electroencephalogram, and electrooculogram for sleep staging and electromyogram. Another solution for the diagnosis is the home-based polygraphy [18–20].

CPAP (continuous positive airway pressure) is the gold standard treatment for severe OSA, but since the 1980s, even mandibular advancement devices (MADs) were used as a therapy for OSA. MADs are widely used for mild and moderate OSA treatment, and several studies have demonstrated their effectiveness even in the treatment of severe OSA [21,22]. MADs present higher compliance and are more convenient for the patient than CPAP [23–27].

Although mandibular advancement devices represent the gold standard in the treatment of mild and moderate OSA, they cause skeletal and dental modifications due to their use [28].

MADs are responsible for a small but statistically significant change in the dentition of long-term wearers. Skeletal changes are generally secondary to dental changes. Major dental modifications feature the tendency towards a reduction of overbite and overjet, maxillary incisors palatal tipping and mandibular incisor labial tipping. A moderate linkage exists between the length of treatment time (how long the device is worn), and the number of dental changes experienced [29].

Even if the modifications are not significant compared to the benefits that the use of the MAD brings, it is crucial to evaluate the periodontal consequences of these devices. Periodontal health is defined to be a state free from inflammatory periodontal disease. This situation, in turn, means that absence of inflammation associated with gingivitis or periodontitis [30]. An analysis of the literature reveals an epidemiological picture of the uneven periodontal disease with methodological discrepancies that make it difficult to compare the prevalence and severity of data [31].

In 2007, the Centers for Disease Control and Prevention and the American Academy of Periodontology determined that population-based epidemiological studies should use the definition of moderate and severe periodontitis proposed by Page and Eke [32]. In the European area, the prevalence of moderate periodontitis is between 33.3% and 50%, while the severe one is between 17.6% and 35% [33–35]. Periodontal disease is a prevalent problem; it is estimated that more than 46% of the population in the United States is affected. The advanced forms of periodontitis that result in severe loss of supporting structures and substantial tooth loss affect 8.9% of the population [36].

The American Academy of Sleep Medicine stated that edentulism and severe periodontal disease represent a counter indication to the use of MADs. However, it is not verified how the stresses related to wearing MADs are discharged on the periodontal ligament of teeth [37].

Ethical or instrumental limits and increased complexity often put a limitation to research, and many questions cannot be answered given the impossibility of obtaining satisfactory results. In these cases, reverse engineering (RE) with the finite element method (FEM) represents a solution. RE is a technique used to generate a 3D virtual model from a real-world tangible object. FEM is an engineering instrument used to calculate stress and deformations in complex structures and has been widely applied in biomedical research. In the field of structural engineering, the use of FEM aims to establish the state of tension and deformation of a solid subjected to external actions.

By taking advantage of this engineering resource in orthodontics, it is possible to model and analyze any dental and maxillofacial material or structure [38]. The FEM principle is based on the division of complex structures into smaller sections called elements in which physical properties are applied to study the response of the object to external stimulation such as an orthodontic force. All this represents a significant advantage since the degree of simplification can be controlled.

The study aims to analyze how different types of MADs stress and deform teeth and periodontal ligaments of the individual dental elements of both arches, using a finite element method analysis. To the author's knowledge, this is the first work that compares the effects of three different types of MADs through a FEM. The predictable behavior of teeth and periodontal ligaments (PDLs) suggests to the clinicians the choice of the most suitable OSA treatment device depending on the patient's dental condition.

2. Materials and Methods

The finite element model employed in this study included the maxilla and the mandible with associated teeth and periodontal ligaments and a mandibular advancement device (Figure 1). The model was developed according to the dental cone beam computed tomography (CBCT) (Voxel size 75 micron, FOV 11 × 13 cm. NewTom Giano, Cefla, Italy) images of a 29-year-old female patient. The patient is in good health and does not suffer from OSA or mandibular disorders. Before proceeding with the CT scan, the patient was informed of the purpose of the study and gave her consent to the procedure. Her data were used anonymously in the simulations. The volumetric image of the anatomy was reconstructed in Mimics (v.12.11, Materialise NV, Leuven, Belgium) using an appropriate threshold based on the Hounsfield unit. The periodontal ligaments (PDLs) were reconstructed through the 3D modelling software Rhinoceros (v.5.0 by McNeel & Associates, Seattle, WA) since soft tissues are not identifiable in CT scans. Although the average thickness of periodontal ligament is 0.15 to 0.38 mm and it has an hourglass shape (it is thinnest at the middle third of the root [39]), PDLs were modelled by offsetting each tooth root surface of 0.3 mm to fill the space between each tooth and the alveolar socket [38,40,41]. The devices chosen are among the most used in OSA therapy. The authors have selected to consider two different types of mechanics: Lateral and anterior propulsion. In Herbst's^{TM} case, two lateral telescopic arms protrude the mandible. Somnodent^{TM} also has a lateral mechanism, but the propulsion is no longer given by arms but by two screws that push the wings of the lower splint forward as they move forward. Orthoapnea^{TM} is an entirely different device that has an anterior inverse connecting rod that is activated through the central screw that produces the protrusion of the mandible. In the simulations, Somnodent^{TM} was reproduced without vertical elastics that are used to improve the outcome of the therapy. To simulate the effect of different MAD models, three MADs were created through a RE approach starting from the physical prototype due to the personalised nature. An optical laser scanner (Konica Minolta Range 7) was used for digitising a model of MAD to be used as a reference for the design of the two other devices. All the triangular mesh models were further processed and reconstructed to obtain the non-uniform rational basis spline (NURBS) models ready to be imported into Ansys v.19 R1 (Canonsburg, PA, USA) for the final arrangement of the model in terms of material properties, meshing, boundary conditions, and loading conditions.

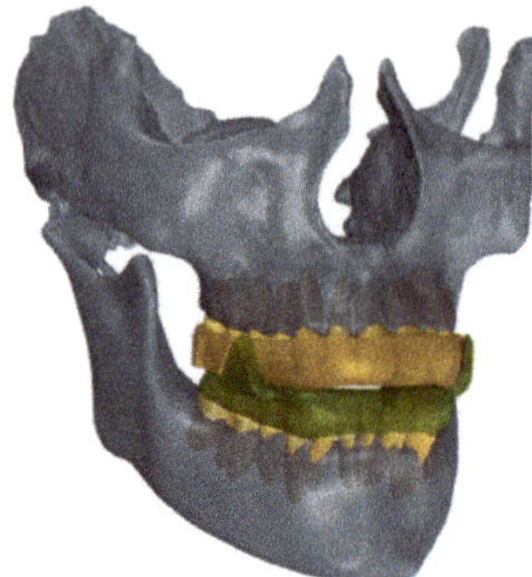

Figure 1. 3D finite element model.

The material properties of teeth and MAD as presented in Table 1 were assumed to be linear elastic, homogeneous, and isotropic. At the same time, a hyperelastic law was used to simulate large non-linear strains associated with the non-linear nature of PDLs [42].

Table 1. Material properties used in the finite element model.

Material	Young's Modulus [MPa]	Poisson's Ratio
Acrylic MAD	8300	0.28
Tooth	18,300	0.31

A 2nd-order Ogden model was used to define the strain energy function [43–45]. The parameter values listed in Table 2 were obtained through the fitting of uniaxial experimental data by Natali et al. [46].

Table 2. Parameters of the second-order Ogden model.

Model	Parameter	Value
Hyperelastic-Ogden	α_1	−3.4761
	α_2	18.679
	μ_1	−0.034004
	μ_2	0.00088691
	d_1	0
	d_1	0

Perfect bonding was assumed between the bodies since MADs are anchored on teeth to force the mandible to a protruded position, and PDLs strongly bind the tooth root to the supporting alveolar bone. Thus, a continuous mesh (shared topology) has been created between MAD and teeth and teeth and PDLs with common nodes at the boundary between the parts. The contact between the bone and the PDL was set as bonded meaning that no sliding or separation is allowed. The two splints of the MAD were coupled in a configuration that will enable mutual sliding without separation. The bony components were assumed as rigid bodies since they have more rigid behaviour than PDLs. Moreover, they were constrained in all directions to prevent rigid body motion.

The authors decided to not differentiate bone in its cortical and cancellous components since the main objective of the study is to compare the results in the three devices. The results, therefore, must be understood as comparative and not in absolute terms.

As a boundary condition, a load associated with the force applied by the MAD to protrude the mandible was applied in the connecting region between the two MAD splints. According to the methodology developed by Bruno et al. [47], it was possible to correlate the amount of advancement of the mandible due to the MAD settings to the magnitude of the force exerted by the device and experimentally determined by Cohen-Levy et al. [48]. The application of a pressure transducer to a MAD allowed measuring the force produced by a progressive mandibular advancement. The advancement to which the MADs have been subjected is 9.5 mm, a reasonable value for the treatment of OSA [38]. Thus, an overall force of 11.18 N was applied on each splint in correspondence of the connection mechanism. The 3D models were then discretised in 5,705,381 linear tetrahedral elements with three degrees of freedom per node. A mean mesh size of 1 mm was used to discretize the model. A mesh convergence study was performed to ensure the adequacy of the results from the simulation. The obtained numerical solution will tend toward a unique value of Von Mises stress by increasing the mesh density in the PDLs from 0.26 to 0.154 mm. The physical memory of the workstation did not support further mesh refinement (0.118 mm). Thus, according to the values provided in Table 3 an element size of 0.2 mm was considered an acceptable trade-off between the quality of the mesh and the computational effort of the workstation. A dense and tight mesh was obtained in the thickness of the bodies ensuring at least two elements in the ligaments' wall.

Table 3. Mesh convergence study.

Mesh Size (mm)	Nodes/Elements	Von Mises Stress (MPa)	Elapsed Time (hour)	Memory Used (GB)
0.26	631,362/3,317,272	8.07×10^{-3}	3	18.972
0.20	1,073,911/5,706,491	4.26×10^{-3}	22	33.212
0.15	1,928,454/10,404,873	4.48×10^{-3}	36	70.366

3. Results

In this study, three finite element simulations were carried out by varying the MAD design to analyze the stress distribution and deformation of teeth and PDLs. Tooth movement and pain are the biomechanical response of the stress that develops in teeth and PDLs under the application of a load. Therefore, to evaluate the teeth that are more loaded and the areas that are prone to be compressed and stretched, results in terms of von Mises stress and deformation of teeth and PDLs were presented. (Figure 2).

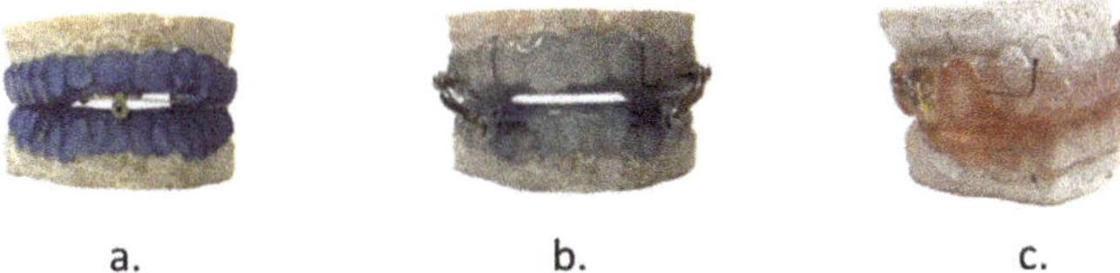

a.   b.   c.

Figure 2. Three models of mandibular advancement devices: OrthoApnea (**a**), Herbst (**b**), Somnodent (**c**).

Regarding the analysis of stress on the periodontal ligaments, significant differences emerge between Orthoapnea™ (Grupo Dental Ortoplus, Malaga, Spain), Herbst™ (Somnomed, Sydney, Australia), and Somnodent™ (Somnomed, Sydney, Australia).

In the case of Orthoapnea™, which has an anterior reverse rod activation mechanism, the stresses on the individual teeth develop mainly on the coronal portion, while apically, the stresses are minor. The most involved teeth are the lower and upper incisors, with the PDLs exhibiting a maximum expression of strength in correspondence of the anterior cervical margin reaching 4.26 kPa (kilopascal). The teeth of the posterior sectors (premolars and molars) are less stressed. The stresses are slightly uneven in the various portions of the arches, as shown in Figure 3.

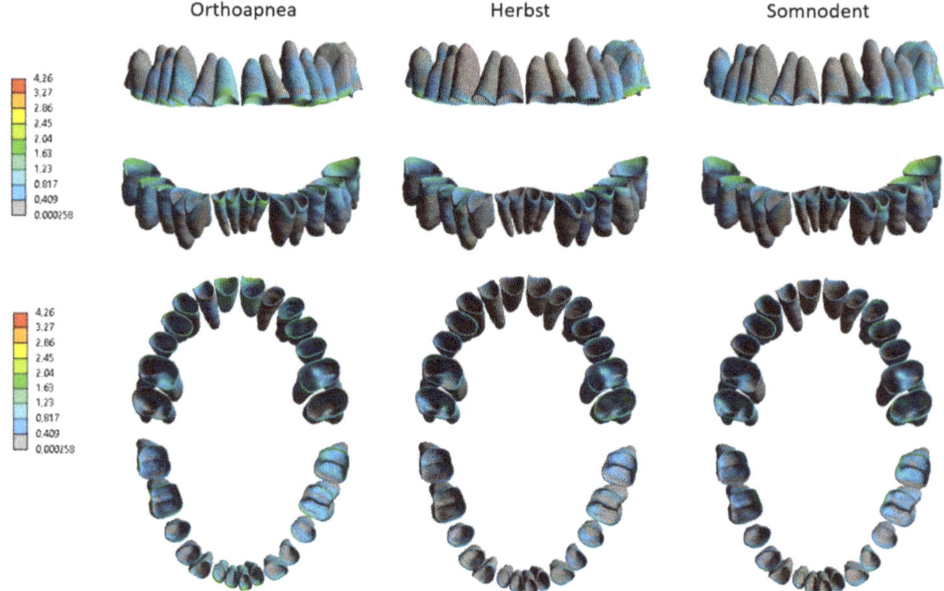

Figure 3. Contour plots. Periodontal ligaments (PDLs) stress distribution (kPa), frontal and occlusal view.

Herbst^{TM} is a bilateral telescopic activation device. It presents periodontal stresses that are concentrated in the most coronal portion of the teeth similarly to Orthoapnea^{mTM}. The main difference with the first device is that tensions caused by Herbst^{TM} are much more uniform in the arches. The most stressed teeth are the upper and lower premolars and molars (but the differences are not significant while compared with other teeth). The PDLs are mainly affected by stress in the medio-distal cervical region with a peak of pressure that reaches 3.56 kPa.

Somnodent^{TM} is activated with two lateral propulsion screws and this case also presents concentrated stresses, especially in the coronal portion of the teeth. Stresses are concentrated in the periodontal ligaments of the second molars and, with a minor component, the anterior teeth of both arches. The stress mainly affects the medio-distal cervical sides of PDLs and the lingual sides of mandibular PDLs associated with molars. The maximum pressure achieved is about 3.27 kPa. Somnodent^{TM} is the device that generates the least stresses among the three considered in the study. In Figure 4, the stress analysis of dental elements reflects periodontal tensions. In Orthoapnea^{TM}, the most loaded teeth are the upper and lower incisors. The most affected dental portions are the incisal margins with a maximum value of 600 kPa. Herbst^{TM} has less intense stresses that are concentrated mainly in the maxillary molars and the mandibular premolars. The maximum load value reached is 302 kPa. Somnodent^{TM} has stresses concentrated on the upper molars and mandibular second premolars in correspondence of the splint's wings. The highest pressure is 287 kPa. The analyses also allowed us to obtain images, from the occlusal view, representing the stresses and deformations on periodontal ligaments (Figure 3) and teeth (Figure 4).

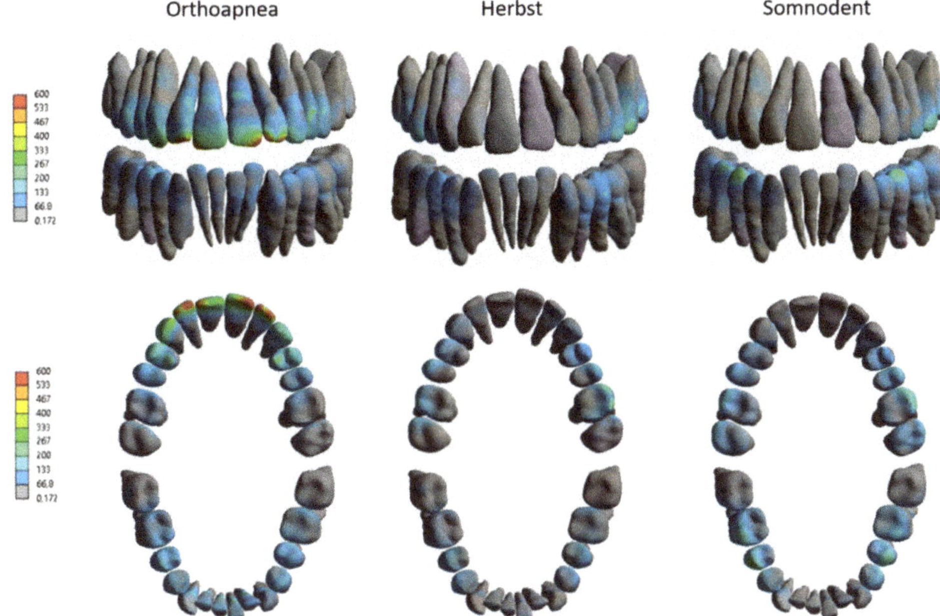

Figure 4. Colour plots. Stress distribution (kPa) in the teeth, frontal and occlusal view.

A lateral representation of the arches shows the deformation vectors to which the individual dental elements of both arches are submitted to (Figure 5). The model shows a right-side view, but the left side is considered symmetrical. In all three devices, the forces tend to displace the teeth of the upper arch downwards and backwards when considering the anterior teeth. In contrast, the deformations on the posterior teeth are directed back and upwards. The deformations are more significant in Orthoapnea^{TM}; in particular, incisor teeth are the most affected teeth, as shown in Figure 5, while the deformations are more uniform in Herbst^{TM} and in particular in Somnodent^{TM}, which is characterized by the lowest intensity of deformation (Figure 5). This image does not report the advancement of the 9.5 mm MAD at which the simulations were made but is only indicative of the direction of displacement.

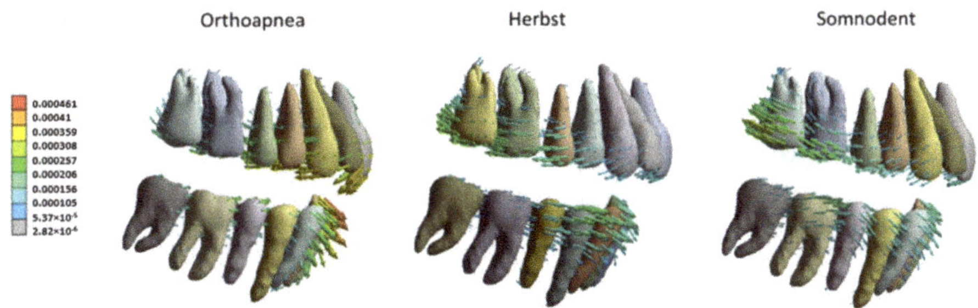

Figure 5. Deformation (mm) on teeth from a lateral view. Arrows indicate the direction of tooth displacement and its intensity (red for higher displacement; blue and grey for lower displacement).

4. Discussion

Mandibular advancement devices are efficient in reducing the apnea-hypopnea index (AHI) under the score of 5, a value below which sporadic apnea episodes are considered physiological, in both mild and severe OSA. The resolution of the apneic syndrome is obtained in 48% of treated patients who presented a mild-severe OSA and AHI under 10 in 64% of the patients, which is considered a good result starting from a severe OSA [18,49]. The reduction of snoring and daily sleepiness, which are the most common symptoms of sleep apnea, is referred in 82% of patients who wears a mandibular advancement device [50–52].

The effectiveness of MAD therapy has been widely verified in the literature. Still, it is essential also to consider the side effects caused by the treatment. The treatment is indicated for the entire duration of the subject's life. Therefore, as this is not a short-term treatment, the consequences at the dental, skeletal, and articular level have to be taken into consideration. Even though a slight mandibular advancement can also be considered a safe procedure for an extended period and should not cause permanent side effects on the temporomandibular joint [51], dental effects are more evident, resulting in a reduction of both overjet and overbite [53].

Dental displacement is the result of the application of the MAD forces on the attachment system of the teeth and the consequent remodeling at the periodontal level. A study that uses a finite element analysis to investigate the optimal orthodontic force establishes that in a range of orthodontic pressure between 4.7 kPa and 16 kPa, dental movements are generated in the biological respect of the periodontal and dental structure [54]. The pressures highlighted by our study turned out to be lower than this range of optimal orthodontic forces. Therefore, a slight tooth movement is obtained, which is compatible with wearing the MAD for long periods. This conclusion is supported by the results of a study showing that occlusal changes (reduction in overjet and overbite) occur in 86.7% of patients [55]. Another recent research evaluating a 10-year follow-up also demonstrates the presence of dental modifications resulting from the use of MAD; these changes are progressive over time [56]. Given the significant prevalence of the periodontal disease, it is crucial to know the stress levels generated at the level of the individual dental elements in the different types of appliance.

Previous studies investigated the effects of forces applied on periodontal ligaments by orthodontic appliances on both healthy and reduced periodontal attachment. It has been demonstrated that the maximum stress value is located in the ligament portion close to the alveolar crest, because of the pressure generated by the mandibular advancement splint on the arch [57–59].

The analysis found that the Somnodent™ and Herbst™ appliances exhibited lower levels of stress than an Orthoapnea™ device. Orthoapnea™ releases the activation forces in a concentrated manner on the anterior teeth of both arches. At the same time, Herbst and Somnodent™ concentrate their stresses on the posterior teeth. This result agrees with a previous study demonstrating a stress concentration at the molars wearing a MAD with the lateral mechanism [38]. The results of the paper agree on a study comparing the dental effects of using a Somnodent™ device and TAPTM (which is a MAD that has the same mechanics as Orthoapnea™). In the study, it was shown that the dental modifications are lower in the case of Somnodent™ [60]. According to the authors, the difference depends on the two different mechanics that promote the activation of the three devices. Somnodent™ has a bilateral activation screw located at the level of the molar teeth, Herbst™ has two telescopic technology pistons that join the two arches bringing the jaw forward. Even in this case, the point of application of the forces is at the latero-posterior level on both sides. Orthoapnea™, on the other hand, has an anterior reverse connecting rod, which discharges the forces on the frontal elements of both arches. The advantages of using a device such as Orthoapnea™ are, in particular, the possibility of having lateral excursions, useful, for example, in the bruxist patient and the ability of the device to force the jaw to advance further if the patient tends to open mouth during sleep. This phenomenon, which is a typical tendence in OSA, decreases the efficiency of MAD (causing mandibular postrotation and retroposition) and, therefore, the effectiveness of the therapy [61,62]. In lateral mechanical devices such as Somnodent™ or Herbst™, since there is no real limit to the opening of the mouth, vertical elastics of adequate length and strength can be used to keep the two parts of the splint in contact. The use of elastics is not necessary in the case of OrthoapneaTM, having included this feature in its mechanics. Devices such as Herbst™ and Somnodent™, on the other hand, have the advantage of being less aggressive with regards to the stresses applied at the periodontal level and therefore represent a better option in the case of periodontally compromised patients in the anterior mandible area.

Since it is not possible to perform an in vivo validation of this study, a comparison of the results with previous studies assessed its reliability. Earlier articles on orthodontic tooth loading and mandibular advancement devices evaluation from both a clinical point [60] of view and an engineering point of view (finite element analysis) [38,57–59] have been taken into account.

A study carried out using FEM simulations has shown that in the case of a reduced periodontium it is necessary to reduce the application of orthodontic forces to generate lower and more uniform stresses that do not worsen the already compromised situation [59]. This result aids practitioners in deciding the type of appliance to be used so that the applied forces are uniform. The forces applied to the MAD do not depend only on the type of device, but also on the degree of activation: Higher activations correspond to higher stresses. In particular, when 70% of the maximum advance is exceeded, the risk of root resorption begins to increase (PDL pressure exceeds 4.7 kPa). A study shows that only below 40% of advancement is safe to avoid root resorption, but this is not a sufficient advancement to obtain the therapeutic effect of MAD [38].

This study provides critical clinical results that may influence the choice of the device. The analysis, however, has some limitations: The stress simulations are static, no load cycles have been performed, so they detach from the model of orthodontic forces applied intermittently. The simulation focuses on the investigation of the response of the periodontal ligament disregarding the effects at the bone level. Since the study is comparative with a focus on the behavior of the three devices, the authors decided to not discriminate bone in its cortical and cancellous components; this represents a limitation of the study, but as future work, the discrimination between cortical and cancellous bone will be considered following the approach proposed by Toniolo et al. [63]. With this improvement, it will be possible to analyze patients suffering from periodontitis (bone resorption and edentulism). The model used to carry out the analyses refers to a young female, without significant temporomandibular or orthodontic problems, not affected by periodontal disease and who has all the dental elements of both arches. In clinical practice, it is difficult to find patients with these characteristics. Therefore, the authors plan to continue the work with analysis on the reduced periodontium and with cases of partial edentulism or implant rehabilitations to make the study more comparable with the cohorts of adult patients with OSA. Other simulations can be performed using different levels of MADs' activation.

5. Conclusions

This work represents a step of more complex work for evaluating the effects caused by a mandibular advancement device for OSA treatment. The study, which is based on FEM, allowed authors to assess stress and deformation caused by MADs on periodontal ligaments and teeth surfaces.

The findings of this study suggest that by changing the design of the mandibular advancement device, there is a significant difference in terms of stress between a bilateral propulsion mechanism and an anterior rod system. In the case of the anterior mechanism model, the stresses are much more concentrated and intense in the anterior area, creating a situation that could be contraindicated in patients presenting periodontal problems. In a subsequent study, stress and deformations applied to a patient with a reduced periodontium and partial edentulism can be studied.

Author Contributions: Conceptualization, G.B., A.d.S., A.M. and M.M.; Investigation, M.C. and F.Z.; Methodology, G.B., A.d.S., M.C., D.L. and M.M.; Supervision, A.G.; Validation, A.G.; Writing—original draft, G.B., A.d.S., Francesca Zalunardo, M.C. and M.M. All authors have read and agreed to the published version of the manuscript.

Funding: This research received no external funding.

Conflicts of Interest: The authors declare no conflict of interest.

References

1. Laratta, C.R.; Ayas, N.T.; Povitz, M.; Pendharkar, S.R. Diagnosis and treatment of obstructive sleep apnea in adults. *CMAJ* **2017**, *189*, 1481–1488. [CrossRef]
2. Mayer, P.; Pépin, J.L.; Bettega, G.; Veale, D.; Ferretti, G.; Deschaux, C.; Lévy, P. Relationship between body mass index, age and upper airway measurements in snorers and sleep apnoea patients. *Eur. Respir. J.* **1996**, *9*, 1–9. [CrossRef] [PubMed]
3. White, D.P. Pathogenesis of obstructive and central sleep apnea. *Am. J. Respir. Crit. Care Med.* **2005**, *172*, 1363–1370. [CrossRef] [PubMed]
4. Horner, R.L. Contributions of passive mechanical loads and active neuromuscular compensation to upper airway collapsibility during sleep. *J. Appl. Physiol.* **2007**, *102*, 510–512. [CrossRef] [PubMed]
5. Destors, M.; Tamisier, R.; Galerneau, L.M.; Lévy, P.; Pepin, J.L. Pathophysiology of obstructive sleep apnea syndrome and its cardiometabolic consequences. *Presse Med.* **2017**, *46*, 395–403. [CrossRef]
6. Schwartz, R.N.; Payne, R.J.; Forest, V.I.; Hier, M.P.; Fanous, A.; Vallée-Gravel, C. The relationship between upper airway collapse and the severity of obstructive sleep apnea syndrome: A chart review. *J. Otolaryngol. Head Neck Surg.* **2015**, *44*, 32. [CrossRef]
7. Peppard, P.E.; Young, T.; Barnet, J.H.; Palta, M.; Hagen, E.W.; Hla, K.M. Increased prevalence of sleep-disordered breathing in adults. *Am. J. Epidemiol.* **2013**, *177*, 6–14. [CrossRef]
8. Heinzer, R.; Vat, S.; Marques-Vidal, P.; Marti-Soler, H.; Andries, D.; Tobback, N.; Mooser, V.; Preisig, M.; Malhotra, A.; Waeber, G.; et al. Prevalence of sleep-disordered breathing in the general population: The HypnoLaus study. *Lancet Respir. Med.* **2015**, *3*, 310–318. [CrossRef]
9. Lurie, A. *Obstructive Sleep Apnea in Adults*; Karger: Basel, Switzerland, 2011.
10. Chowdhuri, S.; Quan, S.F.; Almeida, F.; Ayappa, I.; Batool-Anwar, S.; Budhiraja, R.; Cruse, P.E.; Drager, L.F.; Griss, B.; Marshall, N.; et al. An official american thoracic society research statement: Impact of mild obstructive sleep apnea in adults. *Am. J. Respir. Crit. Care Med.* **2016**, *193*, 37–54. [CrossRef]
11. Fleetham, J.; Ayas, N.; Bradley, D.; Fitzpatrick, M.; Oliver, T.K.; Morrison, D.; Ryan, F.; Series, F.; Skomro, R.; Tsai, W.; et al. Canadian thoracic Society 2011 guideline update: Diagnosis and treatment of sleep disordered breathing. *Can. Respir. J.* **2011**, *18*, 25–47. [CrossRef]
12. Salmina, D.; Ogna, A.; Wuerzner, G.; Heinzer, R.; Ogna, V.F. Arterial hypertension and obstructive sleep apnea syndrome: State of knowledge. *Rev. Med. Suisse* **2019**, *11*, 1620–1624.
13. Gaines, J.; Vgontzas, A.N.; Fernandez-Mendoza, J.; Bixler, E.O. Obstructive sleep apnea and the metabolic syndrome: The road to clinically-meaningful phenotyping, improved prognosis, and personalized treatment. *Sleep Med. Rev.* **2018**, *42*, 211–219. [CrossRef] [PubMed]
14. Amra, B.; Rahmati, B.; Soltaninejad, F.; Feizi, A. Screening questionnaires for obstructive sleep apnea: An updated systematic review. *Oman Med. J.* **2018**, *33*, 184–192. [CrossRef] [PubMed]
15. Nagappa, M.; Liao, P.; Wong, J.; Auckley, D.; Ramachandran, S.K.; Memtsoudis, S.; Mokhlesi, B.; Chung, F. Validation of the stop-bang questionnaire as a screening tool for obstructive sleep apnea among different populations: A systematic review and meta-Analysis. *PLoS ONE* **2015**, *10*, 14–36. [CrossRef] [PubMed]
16. Chiu, H.Y.; Chen, P.Y.; Chuang, L.P.; Chen, N.H.; Tu, Y.K.; Hsieh, Y.J.; Wang, Y.C.; Guilleminault, C. Diagnostic accuracy of the Berlin questionnaire, STOP-BANG, STOP, and Epworth sleepiness scale in detecting obstructive sleep apnea: A bivariate meta-analysis. *Sleep Med. Rev.* **2017**, *36*, 57–70. [CrossRef] [PubMed]
17. Lonia, L.; Scalese, M.; Rossato, G.; Bruno, G.; Zalunardo, F.; De Stefani, A.; Gracco, A. Validity of the stop-bang questionnaire in identifying osa in a dental patient cohort. *Medicina* **2020**, *56*, 324. [CrossRef] [PubMed]
18. Kapur, V.K.; Auckley, D.H.; Chowdhuri, S.; Kuhlmann, D.C.; Mehra, R.; Ramar, K.; Harrod, C.G. Clinical practice guideline for diagnostic testing for adult obstructive sleep apnea: An American academy of sleep medicine clinical practice guideline. *J. Clin. Sleep Med.* **2017**, *13*, 472–477. [CrossRef] [PubMed]
19. De Oliveira, A.C.T.; Martinez, D.; Vasconcelos, L.F.T.; Gonçalves, S.C.; do Carmo Lenz, M.; Fuchs, S.C.; Gus, M.; de Abreu-Silva, E.O.; Moreira, L.B.; Fuchs, F.D. Diagnosis of obstructive sleep apnea syndrome and its outcomes with home portable monitoring. *Chest* **2009**, *135*, 330–336. [CrossRef]
20. Su, S.; Baroody, F.M.; Kohrman, M.; Suskind, D. A comparison of polysomnography and a portable home sleep study in the diagnosis of obstructive sleep apnea syndrome. *Otolaryngol. Head Neck Surg.* **2004**, *131*, 44–50. [CrossRef]

21. De Stefani, A.; Bruno, G.; Agostini, L.; Mezzofranco, L.; Gracco, A. Resolution of a Severe Grade of Obstructive Sleep Apnea Syndrome with Mandibular Advancement Device: A Case Report. *Sleep Med. Res.* **2020**, *11*, 44–48. [CrossRef]

22. Luzzi, V.; Brunori, M.; Terranova, S.; Di Paolo, C.; Ierardo, G.; Vozza, I.; Polimeni, A. Difficult-to-treat OSAS: Combined continuous positive airway pressure (CPAP) and mandibular advancement devices (MADs) therapy. A case report. *Cranio J. Craniomandib. Pract.* **2020**, *38*, 196–200. [CrossRef] [PubMed]

23. Phillips, C.L.; Grunstein, R.R.; Darendeliler, M.A.; Mihailidou, A.S.; Srinivasan, V.K.; Yee, B.J.; Marks, G.B.; Cistulli, P.A. Health outcomes of continuous positive airway pressure versus oral appliance treatment for obstructive sleep apnea: A randomized controlled trial. *Am. J. Respir. Crit. Care Med.* **2013**, *187*, 79–87. [CrossRef] [PubMed]

24. Ferguson, K.A.; Cartwright, R.; Rogers, R.; Schmidt-Nowara, W. Oral appliances for snoring and obstructive sleep apnea: A review. *Sleep* **2006**, *29*, 244–262. [CrossRef] [PubMed]

25. Sutherland, K.; Vanderveken, O.M.; Tsuda, H.; Marklund, M.; Gagnadoux, F.; Kushida, C.A.; Cistulli, P.A. Oral appliance treatment for obstructive sleep apnea: An update. *J. Clin. Sleep Med.* **2014**, *10*, 15–27. [CrossRef]

26. De Stefani, A.; Bruno, G.; Mezzofranco, L.; Perri, A.; Marchese Ragona, R.; Gracco, A. Multidisciplinary ent-orthodontic treatment in a hypertensive patient affected by severe OSAS. *ORAL Implantol.* **2018**, *11*, 59–63.

27. Mickelson, S.A. Oral Appliances for Snoring and Obstructive Sleep Apnea. *Otolaryngol. Clin. N. Am.* **2020**, *53*, 397–407. [CrossRef]

28. Ramar, K.; Dort, L.C.; Katz, S.G.; Lettieri, C.J.; Harrod, C.G.; Thomas, S.M.; Chervin, R.D. Clinical Practice Guideline for the Treatment of Obstructive Sleep Apnea and Snoring with Oral Appliance Therapy: An Update for 2015: An American Academy of Sleep Medicine and American Academy of Dental Sleep Medicine clinical practice guideline. *J. Dent. Sleep Med.* **2015**, *11*, 773–827. [CrossRef]

29. De Martins, O.F.M.; Chaves Junior, C.M.; Rossi, R.R.P.; Cunali, P.A.; Dal-Fabbro, C.; Bittencourt, L. Side effects of mandibular advancement splints for the treatment of snoring and obstructive sleep apnea: A systematic review. *Dent. Press J. Orthod.* **2018**, *23*, 45–54. [CrossRef]

30. Lang, N.P.; Bartold, P.M. Periodontal health. *J. Clin. Periodontol.* **2018**, *89*, 9–16. [CrossRef]

31. Savage, A.; Eaton, K.A.; Moles, D.R.; Needleman, I. A systematic review of definitions of periodontitis and methods that have been used to identify this disease. *J. Clin. Periodontol.* **2009**, *36*, 458–467. [CrossRef]

32. Eke, P.I.; Page, R.C.; Wei, L.; Thornton-Evans, G.; Genco, R.J. Update of the Case Definitions for Population-Based Surveillance of Periodontitis. *J. Periodontol.* **2012**, *78*, 1387–1399. [CrossRef] [PubMed]

33. Holtfreter, B.; Schwahn, C.; Biffar, R.; Kocher, T. Epidemiology of periodontal diseases in the study of health in Pomerania. *J. Clin. Periodontol.* **2009**, *36*, 114–123. [CrossRef] [PubMed]

34. Holtfreter, B.; Kocher, T.; Hoffmann, T.; Desvarieux, M.; Micheelis, W. Prevalence of periodontal disease and treatment demands based on a German dental survey (DMS IV). *J. Clin. Periodontol.* **2010**, *37*, 211–219. [CrossRef] [PubMed]

35. Aimetti, M.; Perotto, S.; Castiglione, A.; Mariani, G.M.; Ferrarotti, F.; Romano, F. Prevalence of periodontitis in an adult population from an urban area in North Italy: Findings from a cross-sectional population-based epidemiological survey. *J. Clin. Periodontol.* **2015**, *42*, 622–631. [CrossRef]

36. Eke, P.I.; Dye, B.A.; Wei, L.; Slade, G.D.; Thornton-Evans, G.O.; Borgnakke, W.S.; Taylor, G.W.; Page, R.C.; Beck, J.D.; Genco, R.J. Update on Prevalence of Periodontitis in Adults in the United States: NHANES 2009 to 2012. *J. Periodontol.* **2015**, *86*, 11–22. [CrossRef]

37. Kushida, C.A.; Morgenthaler, T.I.; Littner, M.R.; Alessi, C.A.; Bailey, D.; Coleman, J., Jr.; Friedman, L.; Hirshkowitz, M.; Kapen, S.; Kramer, M.; et al. Practice parameters for the treatment of snoring and obstructive sleep apnea with oral appliances: An update for 2005. *Sleep* **2006**, *29*, 240–243. [CrossRef]

38. Lee, J.S.; Choi, H.I.; Lee, H.; Ahn, S.J.; Noh, G. Biomechanical effect of mandibular advancement device with different protrusion positions for treatment of obstructive sleep apnoea on tooth and facial bone: A finite element study. *J. Oral Rehabil.* **2018**, *45*, 948–958. [CrossRef]

39. Chandra, S.; Chandra, S.; Chandra, M.; Chandra, N. *Textbook of Dental and Oral Histology with Embryology*; Jaypee Brothers Medical Publishers (P) Ltd.: New Delhi, India, 2007.

40. Dorow, C.; Schneider, J.; Sander, F.G. Finite element simulation of in vivo tooth mobility in comparison with experimental results. *J. Mech. Med. Biol.* **2003**, *3*, 79–94. [CrossRef]

41. Ammar, H.H.; Ngan, P.; Crout, R.J.; Mucino, V.H.; Mukdadi, O.M. Three-dimensional modeling and finite element analysis in treatment planning for orthodontic tooth movement. *Am. J. Orthod. Dentofac. Orthop.* **2011**, *139*, e59–e71. [CrossRef]

42. Liu, Z.; Qian, Y.; Zhang, Y.; Fan, Y. Effects of several temporomandibular disorders on the stress distributions of temporomandibular joint: A finite element analysis. *Comput. Methods Biomech. Biomed. Eng.* **2016**, *19*, 137–143. [CrossRef]

43. Wu, B.; Tang, W.; Yan, B. Study on stress distribution in periodontal ligament of impacted tooth based on hyperelastic model. In Proceedings of the 2009 International Conference on Information Engineering and Computer Science, ICIECS, Wuhan, China, 19–20 December 2009; pp. 1–4.

44. Huang, H.; Tang, W.; Yan, B.; Wu, B. Mechanical responses of periodontal ligament under a realistic orthodontic loading. *Procedia Eng.* **2012**, *31*, 828–833.

45. Roostaie, M.; Soltani, M. Mechanical responses of maxillary canine and surrounding tissues under orthodontic loading: A non-linear three-dimensional finite element analysis. *J. Braz. Soc. Mech. Sci. Eng.* **2017**, *39*, 2353–2369. [CrossRef]

46. Natali, A.N.; Carniel, E.L.; Pavan, P.G.; Bourauel, C.; Ziegler, A.; Keilig, L. Experimental-numerical analysis of minipig's multi-rooted teeth. *J. Biomech.* **2007**, *40*, 1701–1708. [CrossRef] [PubMed]

47. Bruno, G.; De Stefani, A.; Conte, E.; Caragiuli, M.; Mandolini, M.; Landi, D.; Gracco, A. A procedure for analyzing mandible roto-translation induced by mandibular advancement devices. *Materials (Basel)* **2020**, *13*, 1826. [CrossRef] [PubMed]

48. Cohen-Levy, J.; Pételle, B.; Pinguet, J.; Limerat, E.; Fleury, B. Forces created by mandibular advancement devices in OSAS patients: A pilot study during sleep. *Sleep Breath.* **2013**, *17*, 781–789. [CrossRef]

49. Azagra-Calero, E.; Espinar-Escalona, E.; Barrera-Mora, J.M.; Llamas-Carreras, J.M.; Solano-Reina, E. Obstructive sleep apnea syndrome (OSAS). Review of the literature. *Med. Oral Patol. Oral Cir. Bucal* **2012**, *17*, 925–929. [CrossRef]

50. Serra-Torres, S.; Bellot-Arcís, C.; Montiel-Company, J.M.; Marco-Algarra, J.; Almerich-Silla, J.M. Effectiveness of mandibular advancement appliances in treating obstructive sleep apnea syndrome: A systematic review. *Laryngoscope* **2016**, *126*, 507–514. [CrossRef]

51. Crivellin, G.; Bruno, G.; De Stefani, A.; Mazzoli, A.; Mandolini, M.; Brunzini, A.; Gracco, A. Strength distribution on TMJ using mandibular advancement device for OSAS treatment: A finite element study. *Dent. Cadmos* **2018**, *86*, 757–764. [CrossRef]

52. Lu, R.J.; Tian, N.; Wang, J.Z.; Zou, X.; Wang, J.J.; Zhang, M.; Bai, C.Q.; Yu, K.T. The effectiveness of adjustable oral appliance for older adult patients with obstructive sleep apnea syndrome. *Ann. Cardiothorac. Surg.* **2020**, *9*, 2178–2186.

53. Pliska, B.T.; Nam, H.; Chen, H.; Lowe, A.A.; Almeida, F.R. Obstructive sleep apnea and mandibular advancement splints: Occlusal effects and progression of changes associated with a decade of treatment. *J. Clin. Sleep Med.* **2014**, *10*, 1285–1291. [CrossRef]

54. Liao, Z.; Chen, J.; Li, W.; Darendeliler, M.A.; Swain, M.; Li, Q. Biomechanical investigation into the role of the periodontal ligament in optimising orthodontic force: A finite element case study. *Arch. Oral Biol.* **2016**, *66*, 98–107. [CrossRef]

55. Ueda, H.; Almeida, F.R.; Lowe, A.A.; Ruse, N.D. Changes in occlusal contact area during oral appliance therapy assessed on study models. *Angle Orthod.* **2008**, *78*, 866–872. [CrossRef] [PubMed]

56. Uniken Venema, J.A.M.; Doff, M.H.J.; Joffe-Sokolova, D.S.; Wijkstra, P.J.; van der Hoeven, J.H.; Stegenga, B.; Hoekema, A. Dental side effects of long-term obstructive sleep apnea therapy: A 10-year follow-up study. *Clin. Oral Investig.* **2019**, *24*, 3069–3076. [CrossRef] [PubMed]

57. Brunzini, A.; Gracco, A.; Mazzoli, A.; Mandolini, M.; Manieri, S.; Germani, M. Preliminary simulation model toward the study of the effects caused by different mandibular advancement devices in OSAS treatment. *Comput. Methods Biomech. Biomed. Eng.* **2018**, *21*, 693–702. [CrossRef] [PubMed]

58. Toms, S.R.; Eberhardt, A.W. A nonlinear finite element analysis of the periodontal ligament under orthodontic tooth loading. *Am. J. Orthod. Dentofac. Orthop.* **2003**, *123*, 657–665.

59. Jeon, P.D.; Turley, P.K.; Ting, K. Three-dimensional finite element analysis of stress in the periodontal ligament of the maxillary first molar with simulated bone loss. *Am. J. Orthod. Dentofac. Orthop.* **2001**, *119*, 498–504.

60. Venema, J.; Stellingsma, C.; Doff, M.; Hoekema, A. Dental Side Effects of Long-Term Obstructive Sleep Apnea Therapy: A Comparison of Three Therapeutic Modalities. *J. Dent. Sleep Med.* **2018**, *5*, 39–46. [CrossRef]

61. Tsuda, H.; Lowe, A.A.; Chen, H.; Fleetham, J.A.; Ayas, N.T.; Almeida, F.R. The relationship between mouth opening and sleep stage-related sleep disordered breathing. *J. Clin. Sleep Med.* **2011**, *7*, 181–186. [CrossRef]
62. Kim, E.J.; Choi, J.H.; Kim, K.W.; Kim, T.H.; Lee, S.H.; Lee, H.M.; Shin, C.; Lee, K.Y.; Lee, S.H. The impacts of open-mouth breathing on upper airway space in obstructive sleep apnea: 3-D MDCT analysis. *Eur. Arch. Oto-Rhino-Laryngol.* **2011**, *268*, 533–539. [CrossRef]
63. Toniolo, I.; Salmaso, C.; Bruno, G.; De Stefani, A.; Stefanini, C.; Gracco, A.L.T.; Carniel, E.L. Anisotropic computational modelling of bony structures from CT data: An almost automatic procedure. *Comput. Methods Programs Biomed.* **2020**, *189*, 105319.

applied
sciences

Article

Temperature Changes in Composite Materials during Photopolymerization

Leszek Szalewski [1], Magdalena Szalewska [1,2,*], Paweł Jarosz [3], Michał Woś [4] and Jolanta Szymańska [1]

[1] Department of Integrated Paediatric Dentistry, Chair of Integrated Dentistry, Medical University of Lublin, 20-093 Lublin, Poland; Leszek.szalewski@umlub.pl (L.S.); Jolanta.szymanska@umlub.pl (J.S.)
[2] Doctoral School, Medical University of Lublin, 20-093 Lublin, Poland
[3] Engineering Studies Centre, The Institute of Technical Sciences and Aviation, The State School of Higher Education in Chełm, 22-100 Chełm, Poland; pjarosz@pwszchelm.edu.pl
[4] Department of Computer Science and Medical Statistics with the Studio of Remote Learning, Medical University of Lublin, 20-093 Lublin, Poland; 49629@student.umlub.pl
* Correspondence: 35812@student.umlub.pl

Abstract: During polymerization, composite materials cause a temperature rise which may lead to irreversible changes in the dental pulp. The mechanical properties of composite materials depend on a number of factors, such as the composition of the material, the type of polymerization unit, the polymerization mode, and the duration of polymerization. The objective of this study was to assess the temperature rise values and flexural strength of composite materials, as obtained using different modes and times of polymerization. A total of six composite materials were used in the study. Samples of each of the materials were cured using seven polymerization protocols. A CMP-401 digital meter (Sonel, Świdnica, Poland), complete with a type K thermocouple (NiCr-Ni), was used to record the temperature increases during the light curing of the resin composites. Temperature rises were recorded beneath the composite disc in an acrylic matrix. The specimens were tested for flexural strength using a Cometech QC-508M2 testing machine. The lowest results for the increased mean temperature were obtained for Fast-Cure 3 s (39.0 °C), while the highest results were obtained for Fast-Cure 20 s (45.8 °C). The highest average temperature values for all tested protocols were recorded for the Z550 Filtek material. Mean flexural strengths as measured in each test group were higher than the minimum value for composite materials as per the ISO:4049 standard. In the case of deep caries with a thin layer of dentin separating the filling from pulp, a base layer or a short polymerization duration mode is recommended to protect pulp from thermal injury.

Keywords: temperature rise; composites; polymerization; flexural strength

Citation: Szalewski, L.; Szalewska, M.; Jarosz, P.; Woś, M.; Szymańska, J. Temperature Changes in Composite Materials during Photopolymerization. *Appl. Sci.* **2021**, *11*, 474. https://doi.org/10.3390/app11020474

Received: 29 November 2020
Accepted: 2 January 2021
Published: 6 January 2021

Publisher's Note: MDPI stays neutral with regard to jurisdictional claims in published maps and institutional affiliations.

1. Introduction

Composite materials are widely used in every dental practice. As the result of many years of technological development and improvements, as well as changes in their composition, composite materials are very durable, while being esthetically pleasing and popular among dentists. At the same time, numerous manufacturers have introduced increasingly advanced units featuring modifying polymerization programs and power adjustments, so as to ensure the best mechanical performance of composite materials. However, many dentists fail to use polymerization units in the correct manner, by focusing mainly on the anatomical representation and the esthetics of fillings. In the survey carried out by Kopperud et al., almost one third of dentists failed to use proper eye protection against blue light while up to 78.3% of respondents were unaware of the irradiance values of the polymerization lamp they used [1]. Many respondents did not check the quality of the light produced by the polymerization unit. Other studies have shown that preclinical dental students and dentists in their internship years use polymerization lamps in an incorrect manner, not delivering the required amount of energy to the composite layer.

Following a briefing on the use of polymerization tips, the number of study subjects who failed to deliver the minimum required energy to the filling was significantly reduced [2]. Many dentists do not analyze polymerization unit power and modes before purchasing, study instructions of dental composite polymerization protocols, or obey polymerization procedures. This is an essential element, which can affect the final treatment result.

As materials science advanced, more and more dental practitioners abandoned the use of base layers, even in deep caries; combined with the increased lamp powers and changes in composite material compositions, this might lead to increased temperatures being achieved during polymerization, and leading to pulp damage. The critical value for pulp damage (temperature increased by 5.5 °C) was first reported by Zach and Cohen in 1965 on the basis of a study conducted in five adult rhesus macaque monkeys. The authors demonstrated necrosis of 15% of the tissue as the pulp temperature was increased by 5.5 °C [3]. However, not all studies confirmed this value as critical for pulp damage. Gross et al. observed no histological changes within the pulp as its temperature increased by 5.5 °C [4]. In the study by Runnacles et al., a 60-s exposure led to the highest increase in pulp temperatures, exceeding 5.5 °C for certain teeth. However, the authors noted that the critical temperature of pulp damage leading to potential necrosis had been determined by testing the teeth of monkeys rather than humans [5]. There are various methods of protecting the pulp against thermal injury: cooling with water during preparation or using liners, such as calcium hydroxide, mineral trioxide aggregate (MTA), or glass-ionomer cements (GIC) [6–8]. Polymerization lamp choice is essential to avoid pulp overheating. On the other hand, the lamps emitting lower levels of energy can disturb the process of polymerization, thus disabling the acquisition of optimal mechanical parameters, including the tensile strength and the flexural strength.

Since the tensile strength of composites is much lower than their compressive strength, and since tensile strength is typically much more affected by internal flaws, this property is likely the most appropriate test of strength. However, it is usually substituted by measurements of flexural strength as a potentially simpler testing method, well-related to tensile failure. Flexure testing is the standard means for the strength testing of dental composites (ISO 4049), and has been shown to correlate with material wear in some studies [9]. Manufacturers keep upgrading their composite materials using different resins or fillers. Sideridou et al. confirmed that the higher the percentage content of the filler, the higher the flexural strength [10]. The mechanical characteristic of composite materials is affected not only by the composition of materials, but also by the polymerization mode [11–15].

The main objective of this study was to assess the temperature rise values as observed for composite materials during polymerization, and to further assess whether different polymerization durations and modes affect the mechanical properties of materials as exemplified by their flexural strength. The null hypothesis was that the different modes and durations of polymerization would have no effect on the rise of the temperature and flexural strength of the composite materials.

2. Materials and Methods

2.1. Composites

A total of 6 composite materials in A2, Medium Dentin, and Universal shades were used in the study. These included: Essentia: Universal and Medium Dentin (GC Corporation, Tokyo, Japan), GrandioSO, Polofil Supra (VOCO GmbH, Cuxhaven, Germany), Filtek Z550 (3M ESPE, Minneapolis, MN, USA), and Boston (Arkona LFS, Nasutów, Poland). One half of the materials consisted of nano-hybrid materials, while the other half consisted of micro-hybrid materials (Figure 1). The weight content of the filler material in the study group of materials ranged from 76.5% to 89%. All materials featured resin-based matrices: Bis-GMA, TEGDMA, UDMA (with the exception of GrandioSO), and Bis-EMA (with the exception of Polofil Supra); the materials differed by the addition of resins such as Bis-MEPP (Essentia), PEGDMA (Filtek Z550), and HEMA (Polofil Supra). The detailed compositions of the materials used in the study are given in Table 1.

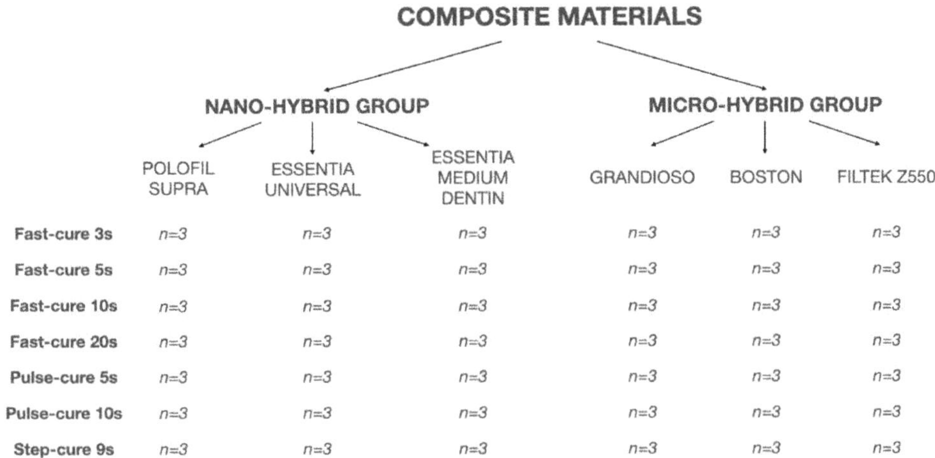

Figure 1. The experimental study design and distribution of samples in groups with curing protocol.

Table 1. Characteristics of composite materials used in the study.

Material	Manufacturer	Shade	Type	Filler Content % (w/w)	Filler Type	Particle Size	Matrix
Boston	Arkona LFS, Nasutów, Poland	A2	nano-hybrid	78%	barium-aluminium-silicon glass, fumed silica, titanium dioxide	15 nm–2000 nm	Bis-GMA, UDMA, Bis-EMA, TEGDMA
Grandioso	Voco, Cuxhaven, Germany	A2	nano-hybrid	89%	glass ceramic filler; silicon dioxide nanoparticles	60% of paticles: 20–40 nm	Bis-GMA, Bis-EMA, TEGDMA
Filtek Z550	3M ESPE, St Paul, MN, USA	A2	nano-hybrid	82%	surface modified zirconia/silica filler, non-agglomerated/non-aggregated surface-modified silica particles	20–3000 nm	Bis-GMA, UDMA, Bis-EMA, PEGDMA, TEGDMA
Essentia	GC Corporation, Tokyo, Japan	Medium Dentin	micro-hybrid	81%	prepolymerized fillers, barium glass, fumed silica	no data	Bis-GMA, UDMA Bis-MEPP, Bis-EMA, TEGDMA
Essentia	GC Corporation, Tokyo, Japan	Universal	micro-hybrid	81%	prepolymerized fillers, barium glass, fumed silica	no data	Bis-GMA, UDMA Bis-MEPP, Bis-EMA, TEGDMA
Polofil Supra	Voco, Cuxhaven, Germany	A2	micro-hybrid	76.50%	Sintraglass multifillers	50–2000 nm	Bis-GMA, UDMA, TEGDMA, HEMA

2.2. Light Curing Unit

The composite materials examined were polymerized using a high-powered LED LCU (Mini LED III Supercharged, Acteon Group, Merignac, France). According to the manufacturer's information, polymerization power 2000 mW/cm^2 when a 7.5-mm diameter tip was used. Three samples of each tested material were cured using 7 polymerization protocols: 4 Fast-Cure modes (full power for 3, 5, 10, and 20 s (double 10-s program), 2 Pulse-Cure modes (5 and 10 shots of 1-s exposures at full power) and 1 Step-Cure mode (soft start with progressive cycle lasting 9 s). In this study, the total energy ranged from

6 J/cm^2 (Fast-Cure, 3 s) to 40 J/cm^2 (Fast-Cure, 20 s). Table 2 shows details pertaining to the LED unit and its polymerization modes.

Table 2. Details of the light polymerization unit and its polymerization modes.

Model	Manufacturer	Wavelength Range	Central Wavelength	Intensity	Fast-Cure Mode	Pulse-Cure Mode	Step-Cure Mode
MINI LED III Supercharged	**Acteon Group (Merignac, France)**	420–480 nm	455–465 nm	2000 mW/cm^2 $\pm$ 10% for an active fiber diameter of 7.5 mm	Full power for 3/5/10 s	5/10 shots of 1 s (full power with emission of 5/10 successive one second flashes with a rest of period of 250 ms between each flash)	6 s progressively and 3 s at full power
					Total energy: 3 s–6 J/cm^2; 5 s–10 J/cm^2; 10 s–20 J/cm^2; 20 s–40 J/cm^2	Total energy: 5 shots–10 J/cm^2; 10 shots–20 J/cm^2	Total energy: 12 J/cm^2

2.3. Temperature Measurement

One hundred and twenty-six specimens were prepared overall, with a total of twenty-one specimens for each composite material, whereby three specimens (n = 3) were polymerized using one of the seven curing modes. All specimens were prepared in an acrylic resin matrix of the same shape (7.5 mm in diameter, 2 mm deep). The tip of the light curing unit touched the composite resin through a protective cover to simulate the situation of filling a tooth within the oral cavity. The scheme of the test stand is shown in Figure 2.

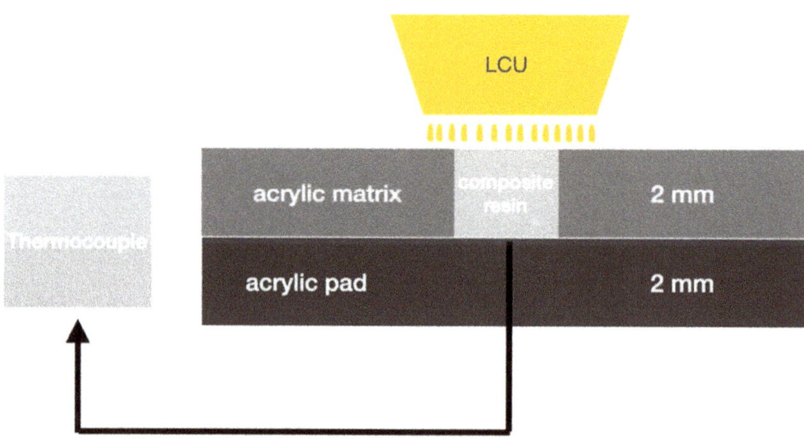

Figure 2. Scheme of the test stand for measuring temperature changes.

All measurements were taken in a temperature-controlled room at a constant temperature of 29 $\pm$ 1 °C. A CMP-401 digital meter (Sonel, Świdnica, Poland) complete with a type K thermocouple (NiCr-Ni) was used to record the temperature increases during the light curing of the resin composites. Temperature rises were recorded beneath the composite disc in the acrylic matrix. Two measurements were recorded during each session, namely the initial temperature and then the maximum temperature achieved in the time of polymerization. The obtained results were analyzed statistically.

2.4. Flexural Strength Test

Seventy specimens were prepared for flexural strength test according to the ISO standard 4049:201012, using the Boston (Arkona LFS, Nasutów, Poland) composite resin (shade A2). Rectangular specimens (25 mm × 2 mm × 2 mm) were produced using a steel mold and placed on a microscope slide to achieve a flat surface. Subsequently, one portion of composite resin was condensed with a dental plugger and flattened by being pressed using another microscope slide. The composite material was then polymerized across a layer of polyethylene film in order to eliminate oxygen inhibition at the surface. Samples were polymerized using a high-powered LED LCU (Mini LED III Supercharged, Acteon Group, Merignac, France) using 7 different modes; the same as for the temperature measurement tests. Ten specimens were used for each mode and duration. Each rectangular sample was polymerized at 4 points. After polymerization, the specimens were released from the mold. Next, the specimens were examined for the presence of air bubbles and defective specimens were excluded from the study. Specimens were then immersed in distilled water at a temperature of 37 °C for 24 h. Next, the specimens were tested for flexural strength using the Cometech QC-508M2 testing machine (Cometech Testing Machines Co., Taichung City, Taiwan) with the opening width of 20 mm, initial gripping force of 1 N and the crosshead speed of 0.75 mm/min. The specimens were measured to an accuracy of 0.01 mm before the test. The test end was marked by the specimen being crushed.

Flexural strength was calculated using the following equation:

$$S = 3FL/(2BH^2)$$

where F is the maximum load in Newtons exerted on the specimens, L is the distance (20 mm) between the supports, accurate to ±0.01 mm, B is the width (2 mm ± 0.01 mm) of the specimens measured immediately prior to testing and H is the height (2 mm ± 0.01 mm) of the specimens measured immediately prior to testing.

2.5. Statistical Analysis

The mean maximum temperatures at the assigned measurement sites and flexural strength were analyzed using ANOVA and Shapiro–Wilk tests. The analyses were conducted using Statistica software version 13 (Statsoft, Warszawa, Poland) at the significance level of 0.05.

3. Results

The specific values of the average temperatures before and during the test along with the standard deviations are summarized in Table 3.

The statistical analysis (Shapiro–Wilk test for normality) showed a significance level for Fast-cure 3 s (W = 0.96364, *p*-value = 0.6732), Fast-cure 5 s (W = 0.95208, *p*-value = 0.4585), Fast-cure 10 s (W = 0.96143, *p*-value = 0.6294), Fast-cure 20 s (W = 0.97366, *p*-value = 0.863), Pulse-cure 5 s (W = 0.96143, *p*-value = 0.6294), Pulse-Cure 10 s (W = 0.95192, *p*-value = 0.4559), and Step-Cure 9 s (W = 0.9568, *p*-value = 0.5412). We cannot reject normality in the test groups. The lowest average results were obtained for Fast-Cure 3 s (39.0 °C ± SD 2.7), while the highest average results were obtained for Fast-Cure 20 s (45.8 °C ± SD 2.0). The temperature increase was the lowest for 3 s of continuous polymerization, amounting to 10.1 °C, while being the highest for 20 s of continuous polymerization (16.3 °C). The highest average temperature values for all tested protocols were recorded for the Z550 Filtek material. The variance estimated based on the variability within the group showed a statistical significance for polymerization time (F = 1.4778) and for the type of filler (F = 2.2404). (Table 4).

Table 3. Mean temperature values and standard deviations (SD) for the composite materials and light curing modes evaluated.

	Fast-Cure 3 s		Fast-Cure 5 s		Fast-Cure 10 s		Fast-Cure 20 s (2 × 10 s)		Pulse-Cure 5 s		Pulse-Cure 10 s		Step-Cure 9 s	
	T0	T1	T0	T1	T0	T1	T0	T1	T0	T1	T0	T1	T0	T1
BOSTON	28.4 (0.3)	36.0 (0.4)	28.3 (0.4)	41.5 (0.6)	28.8 (0.2)	45.7 (0.9)	28.9 (0.3)	46.7 (2.1)	29.3 (0.1)	42.4 (1.8)	29.1 (0.2)	44.8 (1.0)	29.2 (0.1)	45.7 (2.7)
GRANDIOSO	28.6 (0.4)	38.6 (0.8)	28.4 (0.3)	42.3 (1.6)	29.0 (0.4)	41.8 (1.4)	29.5 (0.1)	42.8 (0.6)	29.6 (0.2)	43.4 (1.3)	29.7 (0.1)	44.5 (2.0)	29.8 (0.1)	44.8 (0.4)
FILTEK Z550	29.5 (0.1)	41.6 (1.6)	29.7 (0.3)	48.0 (2.5)	29.8 (0.1)	47.3 (1.4)	29.9 (0.1)	47.7 (1.8)	29.0 (0.1)	44.5 (1.3)	29.8 (0.1)	47.4 (1.8)	29.1 (0.8)	47.2 (1.1)
ESSENTIA MEDIUM DENTIN	29.0 (0.1)	36.3 (1.0)	29.6 (0.2)	42.8 (1.6)	29.7 (0.1)	45.2 (1.3)	30 (0.5)	45.3 (0.7)	29.9 (0.1)	43.9 (2.1)	29.9 (0.2)	43.3 (0.6)	29.3 (0.3)	46.1 (1.8)
ESSENTIA UNIVERSAL	29.0 (0.2)	41.2 (0.3)	28.6 (0.1)	42.6 (2.1)	28.7 (0.2)	43.7 (0.1)	29.3 (0.3)	46.5 (0.7)	29.5 (0.1)	43.3 (0.6)	29.5 (0.3)	44.6 (1.8)	29.6 (0.2)	43.8 (2.6)
POLOFIL SUPRA	28.5 (0.2)	41.1 (0.3)	29.0 (0.2)	46 (1.5)	29.2 (0.2)	47.3 (0.4)	29.5 (0.1)	46.7 (1.1)	29.4 (0.2)	44.6 (2.4)	29.4 (0.1)	45.9 (0.8)	29.6 (0.2)	45.5 (0.5)
TOTAL	28.8 (0.45)	39.0 (2.7) [a]	28.9 (0.6)	43.6 (3.0) [a]	29.2 (0.5)	45.0 (2.3) [b]	29.4 (0.4)	45.8 (2.0) [b]	29.6 (0.2)	43.3 (1.7) [b]	29.6 (0.3)	45.0 (1.9) [b]	29.4 (0.4)	45.2 (1.9) [b]

Standard deviations are in parentheses. Values with identical superscript letters are similar within the same temperatures ($p > 0.05$). T0, beginning temperature; T1, maximum temperature during polymerization.

Table 4. Multiple factor ANOVA test for the time of polymerization and the type of filler.

Response: dt					
	Df	Sum Sq	Mean Sq	F value	Pr(>F)
Filler	1	52.00	52.001	2.2404	0.1371
Polymerization time	6	66.54	11.090	1.4778	0.3237
Residuals	118	2738.89	23.211		

Mean flexural strengths as measured in each test group were higher than the minimum value for composite materials as per the ISO:4049 standard, i.e., 80 MPa (Figure 3).

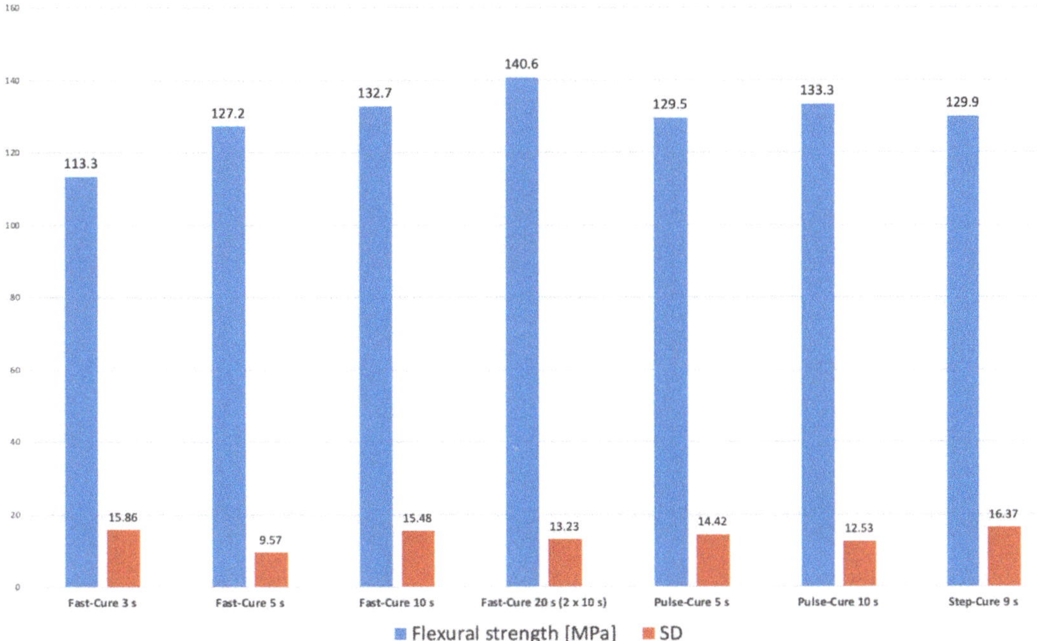

Figure 3. Mean values of flexural strength (MPa) and standard deviations.

The statistical analysis (Shapiro–Wilk test for normality) showed a significance level for Fast-cure 3 s (W = 0.97125, p = 0.90220), Fast-cure 5 s (W = 0.97324 p = 0.91020), Fast-cure 10 s (W = 0.95863, p = 0.86159), Fast-cure 20 s (W = 0.95945, p = 0.90257), Pulse-cure 5 s (W = 0.99865, p = 0.97432), Pulse-cure 10 s (W = 0.96699, p = 0.91351), and Step-cure 9 s (W = 0.95980, p = 0.90619). We cannot reject normality in the test groups. The lowest flexural strength was measured for the Fast-Cure 3 s protocol, while the highest flexural strength was measured for the Fast-Cure 20 s protocol, the differences being statistically significant (p = 0.02060).

4. Discussion

The research results show that the change of the polymerization time has a great impact on the mechanical properties, such as flexural strength. Furthermore, the increase of the material temperature was influenced by the polymerization time. Hence, the null hypothesis was rejected. The results demonstrated that in our study the average temperature as measured for all test materials did not exceed 42.0 °C for Fast-Cure 3 s mode. This value was exceeded for the remaining polymerization modes, reaching a maximum value of 45.8 °C for Fast-Cure 20 s. This value is particularly important for deep cavities with only a thin layer of dentin separating the composite material from the pulp. High temperatures can cause irreversible damage to pulp tissue, and therefore require endodontic treatment.

The study of Khaksaran et al. measured the temperatures following polymerization of bonding systems (N Bond, G-Bond, OptiBond XTR, Clearfil SE, Adper Single Bond 2 and V Bond) on pre-prepared dentin discs obtained from human third molars. The study environment temperature of experiments was 37 °C. The irradiation time was 20 s, and the minimum and maximum temperature rise growth in all subgroups was 1.7 °C and 2.8 °C, respectively. In the case of the 20 s polymerization protocol, no dangerous rise in temperature (5.5 °C) was obtained for either of the bonding systems tested [16]. Our research findings revealed that the maximum temperature increase for the 20 s mode (Fast-cure, 2 × 10 s) was 17.8 °C, for the Filtek Z550 material, with the lowest of 7.25 °C for the Essentia Medium Dentin 3 s mode. In our study the initial temperature was maintained at 29 ± 1 °C. The critical pulp temperature values were not exceeded in the case of Fast-cure 3 s mode. Jo et al., in their study, performed on 30 extracted human molars with class I cavities filled using a nano-hybrid material (Filtek Bulk Fill Posterior Restorative (BFP, 3M ESPE)), found the maximum temperature increase during polymerization at 0.625 mm apically from the top and center of the defect. On the basis of their results, the authors concluded that replacing pulsed or soft start modes with continuous irradiation might reduce the risk of damage to the pulp [17]. The authors' research does not confirm this thesis. For the 5 s modes (Fast-cure and Pulse-cure) and 10 s (Fast-cure, Pulse-cure and Step-cure 9 s) temperature rises were similar: ±14 °C (5 s modes) and ±15.5 °C (10 s modes). Another study (Braga et al.) compared the rise in the temperature during polymerization of two materials (SDR, Dentsply and AURA, SDI) polymerized using two lamps (Bluephase G2, Ivoclar Vivadent, and VALO Cordless, Ultradent) in the standard output power mode. An increase of 6 °C was observed for the Bluephase G2 lamp, as compared to 4 °C for the VALO Cordless lamp for the light curing adhesive system (20 s mode without microcirculation) [18]. Our study demonstrated that for fast-cure 20 s mode (2 × 10 s) the highest temperature increase was 17.8 °C, in the case of Filtek Z550 material. There was no simulated pulpal microcirculation. On the other hand, Kim et al. measured the temperature rise in class I cavities in third molars in the course of layered filling with a composite material (Filtek Z250, Shade A2, lot N506344, 3M ESPE, St. Paul, MN, USA). The temperature was measured for 110 s using eight thermocouples. The authors demonstrated that the rise in the temperature within the cavity was significantly higher during polymerization of the first layer of material (59.8 °C) compared to the next layer (58.5 °C) [19]. The authors' findings do not demonstrate such a temperature increase, even in 20 s mode. This might support the idea of shorter polymerization durations being used for the deepest layers of composite fillings. In a subsequent examination of six composites (Admira, Filtek P60, Premise, Tetric Flow, Tetric Ceram, and Filtek Z250) polymerized using different modes (standard (10 s, full power), pulsed (10 consecutive one-second exposures at full power) or soft start (progressive cycle lasting 20 s)), with a type L thermocouple being used for temperature measurements, higher temperature rises were observed for soft start exposures (in the case of Admira and Tetric Flow materials). The lowest rise in temperature was observed for the Premise material irradiated using the pulsed protocol. None of the exposure protocols tested resulted in temperature being raised to the critical value [20]. In the authors' research the lowest temperature rise was observed for Essentia Medium Dentin and Fast-cure 3 s mode (7.25 °C). No significant differences in temperature rise were found between modes with similar durations. In another study, three polymerization units were compared, one halogen lamp (QTH LCU XL2500 (3M/ESPE), two LED lamps (Freelight LED LCU (3M/ESPE), and Ultrablue (DMC Equipamentos LTDA) on a one composite material (Filtek Z250, 3M/ESPE). Five types of photoactivation modes were used: 20 s with each of the three light curing units according to the manufacturer's instructions, 50 s with the Freelight LED lamp, and 30 s with the Ultrablue IS lamp. The authors demonstrated that for photoactivation times as per the manufacturer's recommendations, both LED lamps produced a lower temperature rise than the QTH lamp (average temperature rise values in degrees Celsius: Ultrablue LED 1.13 (0.05), Freelight LED 1.05 (0.16) vs. QTH XL 2500 1.57 (0.13)). However, the authors stressed that the choice of the type of the polymerization lamp affected the average temperature rises [21]. However, other studies did not confirm the reports suggesting that LED lamps generated lower temperature rise values. A study using 96 fragments of bovine teeth revealed that higher temperatures were obtained during polymerization of composite materials using a LED lamp as compared to halogen lamps. However, the temperature increase was above 5.5 °C for both tested lamps, which could be considered as a critical value [22]. The authors' study demonstrated that the temperature rise was lower than the critical temperature value only for the 3 s Fast-cure mode. For other modes and durations, the final temperature was higher than the critical temperature value. Santini et al. used two LED lamps (Bluephase, Elipar Freelight 2) and a halogen lamp (Prismatics). The authors demonstrated that for all light curing units, the exposure of the bonding system resulted in temperature rises being significantly higher than those observed for the exposure of composite materials. However, higher temperatures were achieved during the polymerization of bonding systems and resin-based composite materials using both LED lamps compared to the halogen lamp [23]. Bagis et al. demonstrated that for the output values of all units tested (halogen, plasma, LED), the temperature rise exceeded 5.5 °C; and the temperature rose along with increasing polymerization duration [24].

Type of polymerization unit and irradiation modes also affect the mechanical properties of composite materials. One such parameter is flexural strength. The study by Pieniak et al. revealed no impact on flexural strength when a LED lamp was used instead of a halogen lamp to polymerize the Filtek Silorane (3M ESPE) and Herculite XRV (Kerr) composite materials. However, a rise in flexural strength was demonstrated as being due to increased polymerization duration when using a halogen lamp for polymerization of Filtek Silorane. In the study, lower values of flexural strength were obtained for methacrylate resin-based materials (Gradia Direct Anterior and Gradia Direct Posterior, GC Japan) [25]. According to our findings, polymerization time affected flexural strength, not the mode of light curing. Another study revealed higher flexural strength values for hybrid composite materials compared to nanofill resins. The authors used two composite materials with different filler types, namely the Filtek P60 3M ESPE (hybrid) and Filtek Supreme 3M ESPE (nanofill) [26]. Another study using the Z250 (3M ESPE) composite and the Scotchbond Multi-purpose Plus (3M ESPE) bonding system revealed that the photoactivation method applied had no impact on the performance of the composite material (including flexural strength), regardless of the material storage medium (water vs. ethanol) [27]. The Filtek Z250 (3M ESPE) and Heliomolar (Ivoclar Vivadent) composite materials were tested by Calheiros et al. at polymerization energy doses of 6–24 J/cm^2. A rise in flexural strength was observed for increasing energy doses for the Filtek Z250 material, whereas no changes in mechanical properties were observed for the Heliomolar composite [28]. The authors' results confirmed the hypothesis that the polymerization energy dose influences the flexural strength value.

All in vitro studies have their limitations. The baseline test stand temperature in our study was lower than the temperature within the cavity. However, even starting at lower temperatures, some values exceeded the limit value of 42 °C. The effects of thermal absorption within the dentin tissue, potentially leading to lower values during polymerization, were not taken into account in the tests. In addition, in the case of deep cavities, it is not always possible to reach the composite layer with the polymerization tip, and the energy dose, leading to the rise in the temperature, is reduced with increasing distance. The effect of the bonding system on the temperature rise was not taken into account in our study, so as to maintain constant conditions for all materials for which different bonding systems are recommended by their manufacturers.

Subsequent studies should be based on the model of human teeth, in which natural heat dissipation in hard tissues will be obtained. In addition, it would be advisable to examine how liners and their thickness affect the transmission of thermal energy to the tooth pulp. Such studies would allow obtaining a safe clinical procedure for the treatment of deep carious lesions.

5. Conclusions

Based on the results of this in vitro study, the following conclusions were drawn:

1. Short polymerization durations lead to lower temperature rise values, while extended polymerization durations increase the values to a critical temperature for pulp damage.
2. The temperature rise was determined by polymerization durations rather than by the exposure mode used.
3. The temperature rise varied according to the materials used.
4. Sufficient flexural strength was obtained for all polymerization modes.

In the case of deep caries with a thin layer of dentin separating the filling from pulp, a base layer or a short duration polymerization mode is recommended to protect pulp from thermal injury during polymerization of the composite materials.

Author Contributions: Conceptualization, L.S., M.S. and J.S.; methodology, L.S. and P.J.; software, M.W.; formal analysis, M.S. and J.S.; investigation, L.S. and P.J.; resources, J.S.; data curation, P.J. and M.W.; writing—original draft preparation, L.S. and M.S.; writing—review and editing, P.J., M.W. and J.S.; visualization, L.S. and M.W.; supervision, J.S.; project administration, M.S. All authors have read and agreed to the published version of the manuscript.

Funding: This research received no external funding.

Data Availability Statement: The data that support the findings of this study are available from the corresponding author, [MS], upon reasonable request.

Conflicts of Interest: The authors declare no conflict of interest.

References

1. Kopperud, S.E.; Rukke, H.V.; Kopperud, H.M.; Bruzell, E.M. Light curing procedures–performance, knowledge level and safety awareness among dentists. *J. Dent.* **2017**, *58*, 67–73. [CrossRef]
2. Suliman, A.A.; Abdo, A.A.; Elmasmari, H.A. Training and experience effect on light-curing efficiency by dental practitioners: Training and experience effect on light-curing. *J. Dent. Educ.* **2020**, *84*, 652–659. [CrossRef] [PubMed]
3. Zach, L.; Cohen, G. Pulp response to externally applied heat. *Oral Surg. Oral Med. Oral Pathol.* **1965**, *19*, 515–530. [CrossRef]
4. Gross, D.J.; Dávila-Sánchez, A.; Runnacles, P.; Zarpellon, D.C.; Kiratcz, F.; Campagnoli, E.B.; Alegría-Acevedo, L.F.; Coelho, U.; Rueggeberg, F.A.; Arrais, C.A.G. In vivo temperature rise and acute inflammatory response in anesthetized human pulp tissue of premolars having Class V preparations after exposure to Polywave® LED light curing units. *Dent. Mater.* **2020**, *36*, 1201–1213. [CrossRef] [PubMed]
5. Runnacles, P.; Arrais, C.A.G.; Pochapski, M.T.; dos Santos, F.A.; Coelho, U.; Gomes, J.C.; De Goes, M.F.; Gomes, O.M.M.; Rueggeberg, F.A. In vivo temperature rise in anesthetized human pulp during exposure to a polywave LED light curing unit. *Dent. Mater.* **2015**, *31*, 505–513. [CrossRef] [PubMed]
6. Garrocho-Rangel, A.; Esparza-Villalpando, V.; Pozos-Guillen, A. Outcomes of direct pulp capping in vital primary teeth with cariously and non-cariously exposed pulp: A systematic review. *Int. J. Paediatr. Dent.* **2020**, *30*, 536–546. [CrossRef] [PubMed]
7. Cervino, G.; Laino, L.; D'Amico, C.; Russo, D.; Nucci, L.; Amoroso, G.; Gorassini, F.; Tepedino, M.; Terranova, A.; Gambino, D.; et al. Mineral Trioxide Aggregate Applications in Endodontics: A Review. *Eur. J. Dent.* **2020**, *14*, 683–691. [CrossRef]
8. Bjørndal, L.; Simon, S.; Tomson, P.L.; Duncan, H.F. Management of deep caries and the exposed pulp. *Int. Endod. J.* **2019**, *52*, 949–973. [CrossRef]
9. Peutzfeldt, A.; Asmussen, E. Modulus of resilience as predictor for clinical wear of restorative resins. *Dent. Mater.* **1992**, *8*, 146–148. [CrossRef]
10. Sideridou, I.D.; Karabela, M.M.; Vouvoudi, E.C. Physical properties of current dental nanohybrid and nanofill light-cured resin composites. *Dent. Mater.* **2011**, *27*, 598–607. [CrossRef]
11. Ilie, N.; Stark, K. Effect of different curing protocols on the mechanical properties of low-viscosity bulk-fill composites. *Clin. Oral Investig.* **2015**, *19*, 271–279. [CrossRef] [PubMed]

12. Aljosa, I.; Tijana, L.; Larisa, B.; Marko, V. Influence of Light-curing Mode on the Mechanical Properties of Dental Resin Nanocomposites. *Procedia Eng.* **2014**, *69*, 921–930. [CrossRef]
13. Catelan, A.; Santo, M.R.d.E.; Menegazzo, L.M.; Moraes, J.C.S.; dos Santos, P.H. Effect of light curing modes on mechanical properties of direct and indirect composites. *Acta Odontol. Scand.* **2013**, *71*, 697–702. [CrossRef] [PubMed]
14. Akiba, S.; Takamizawa, T.; Tsujimoto, A.; Moritake, N.; Ishii, R.; Barkmeier, W.W.; Latta, M.A.; Miyazaki, M. Influence of different curing modes on flexural properties, fracture toughness, and wear behavior of dual-cure provisional resin-based composites. *Dent. Mater. J.* **2019**, *38*, 728–737. [CrossRef] [PubMed]
15. Ajaj, R.; Yousef, M.; Abo El Naga, A. Effect of different curing modes on the degree of conversion and the microhardness of different composite restorations. *Dent Hypotheses* **2015**, *6*, 109. [CrossRef]
16. Khaksaran, N.K.; Kashi, T.J.; Rakhshan, V.; Zeynolabedin, Z.S.; Bagheri, H. Kinetics of pulpal temperature rise during light curing of 6 bonding agents from different generations, using light emitting diode and quartz-tungsten-halogen units: An in-vitro simulation. *Dent. Res. J. (Isfahan)* **2015**, *12*, 173–180.
17. Jo, S.-A.; Lee, C.-H.; Kim, M.-J.; Ferracane, J.; Lee, I.-B. Effect of pulse-width-modulated LED light on the temperature change of composite in tooth cavities. *Dent. Mater.* **2019**, *35*, 554–563. [CrossRef]
18. Braga, S.; Oliveira, L.; Ribeiro, M.; Vilela, A.; da Silva, G.; Price, R.; Soares, C. Effect of Simulated Pulpal Microcirculation on Temperature When Light Curing Bulk Fill Composites. *Oper. Dent.* **2019**, *44*, 289–301. [CrossRef]
19. Kim, R.J.-Y.; Lee, I.-B.; Yoo, J.-Y.; Park, S.-J.; Kim, S.-Y.; Yi, Y.-A.; Hwang, J.-Y.; Seo, D.-G. Real-Time Analysis of Temperature Changes in Composite Increments and Pulp Chamber during Photopolymerization. *BioMed Res. Int.* **2015**, *2015*, 1–6. [CrossRef]
20. Hubbezoglu, I.; Dogan, A.; Dogan, O.M.; Bolayir, G.; Bek, B. Effects of Light Curing Modes and Resin Composites on Temperature Rise under Human Dentin: An in vitro Study. *Dent. Mater. J.* **2008**, *27*, 581–589. [CrossRef]
21. Schneider, L.F.; Consani, S.; Correr-Sobrinho, L.; Correr, A.B.; Sinhoreti, M.A. Halogen and LED light curing of composite: Temperature increase and Knoop hardness. *Clin. Oral Investig.* **2006**, *10*, 66–71. [CrossRef] [PubMed]
22. Guenka Palma-Dibb, R.; Savaris, C.; Alexandra Chinelatti, M.; Augusto de Lima, F.; Bachmann, L.; Jendiroba Faraoni, J. Composite Photopolymerization: Temperature Increase According to Light Source and Dentin Thickness. *JDOI* **2016**, *1*, 11–19. [CrossRef]
23. Santini, A.; Watterson, C.; Miletic, V. Temperature Rise Within the Pulp Chamber During Composite Resin Polymerisation Using Three Different Light Sources. *TODENTJ* **2008**, *2*, 137–141. [CrossRef] [PubMed]
24. Bağiş, B.; Bagis, Y.; Ertas, E.; Ustaomer, S. Comparison of the Heat Generation of Light Curing Units. *J. Contemp. Dent. Pract.* **2008**, *9*, 65–72. [CrossRef]
25. Pieniak, D.; Niewczas, A.M.; Walczak, M.; Zamościńska, J. Influence of photopolymerization parameters on the mechanical properties of polymer-ceramic composites applied in the conservative dentistry. *Acta Bioeng. Biomech.* **2014**, *16*, 29–35. [CrossRef]
26. Da Silva, E.M.; Poskus, L.T.; Guimarães, J.G.A. Influence of Light-polymerization Modes on the Degree of Conversion and Mechanical Properties of Resin Composites: A Comparative Analysis Between a Hybrid and a Nanofilled Composite. *Oper. Dent.* **2008**, *33*, 287–293. [CrossRef]
27. Witzel, M.F.; Calheiros, F.C.; Gonçalves, F.; Kawano, Y.; Braga, R.R. Influence of photoactivation method on conversion, mechanical properties, degradation in ethanol and contraction stress of resin-based materials. *J. Dent.* **2005**, *33*, 773–779. [CrossRef]
28. Calheiros, F.; Kawano, Y.; Stansbury, J.; Braga, R. Influence of radiant exposure on contraction stress, degree of conversion and mechanical properties of resin composites. *Dent. Mater.* **2006**, *22*, 799–803. [CrossRef]

 *applied sciences*

Article

Anisotropic Yield Criterion of Rolled AZ31 Magnesium Alloy via Nanoindentation

Zai Wang [1], Xin Hao [1], Ji Qiu [2], Tao Jin [1,3], Xuefeng Shu [1,2,*] and Xin Li [4,*]

[1]   Institute of Applied Mechanics, College of Mechanical and Vehicle Engineering,
      Taiyuan University of Technology, Taiyuan 030024, China; wangzaityut@163.com (Z.W.);
      haoxin0152@link.tyut.edu.cn (X.H.); jintao@tyut.edu.cn (T.J.)
[2]   College of Aeronautics and Astronautics, Taiyuan University of Technology, Taiyuan 030024, China;
      qiuji@tyut.edu.cn
[3]   Shanxi Key Laboratory of Material Strength and Structural Impact,
      College of Mechanical and Vehicle Engineering, Taiyuan University of Technology, Taiyuan 030024, China
[4]   National Laboratory of Solid State Microstructures, College of Engineering and Applied Science,
      Nanjing University, Nanjing 210093, China
*   Correspondence: shuxuefeng@tyut.edu.cn (X.S.); lixin@nju.edu.cn (X.L.)

Received: 20 November 2020; Accepted: 11 December 2020; Published: 16 December 2020

Abstract: In this paper, the anisotropic mechanical properties of rolled AZ31 magnesium alloys are investigated using nanoindentation tests at room temperature. Nanoindentation was carried out at four angles, including the rolling direction (0°), diagonal direction (45°), transverse direction (90°), and vertical direction (ND). Experimental results show that hardness increases as the rolling angle increases from 0° to 90° and is lowest in the ND direction. The hardness independent of the effect of indentation depth is obtained by analyzing the indentation size effect and then converting hardness values into yield strengths. A new criterion is proposed on the basis of the Hill48 yield criterion. The data obtained through the above experiments are used to determine the parameters in the new criterion. Finally, a solution to the challenge of modeling a function that accurately describes the anisotropic yielding behavior of AZ31 magnesium alloys is proposed using the nanoindentation technique to solve the requirements of specimen size and experimental methods of the macro test.

Keywords: AZ31 magnesium alloy; nanoindentation; indentation size effect; anisotropic yielding criterion

1. Introduction

Magnesium alloys have aroused great research interest and are used in many applications on account of their excellent properties, which include high strength and good wear resistance, corrosion resistance, and thermal stability. However, although magnesium alloys present many advantages over other alloys, they are not perfect metals, because of their poor formability and corrosion resistance at ambient temperature. Magnesium alloys have a hexagonal closed-packed crystal structure with a limited number of available slip systems (basal {0001}, prismatic {1010}, and pyramidal {1011}), which leads to their poor formability during cold working [1]. While cast magnesium alloys are used more extensively than deformed magnesium alloys, the strength, ductility, and mechanical properties of the former are poorer than those of the latter. AZ31 is one of the most widely used alloys currently available. Rolling is an important means to improve the properties of this type of alloy. Under annealing and mechanical twinning, magnesium can show anisotropy at room temperature; thus, the nature of the plastic deformation of magnesium is quite complex [2]. Studies on the deformation and damage behavior of magnesium indicate that anisotropy clearly occurs during magnesium alloy deformation. Initial textures can impact the shape and stress–strain behavior of the samples [1]. The anisotropic mechanical characteristics of AZ31 magnesium alloys have a significant influence on their plastic deformation. Therefore, obtaining a comprehensive understanding of the anisotropic mechanical behavior of magnesium alloys is important.

An accurate description of the magnesium alloy yielding behavior is essential to predict its forming processes. Industrial applications require an overall understanding of the mechanical properties of rolled AZ31 alloys. An accurate mathematical description is significant for predicting material deformation and yielding behaviors; thus, researchers have exerted considerable efforts over the last several decades to establish a solid foundation through which the magnesium yielding and deformation behaviors may be described. In 1864, Tresca established a phenomenological model to describe the yield of materials using the concept of maximum shear stress. Hu proposed an anisotropic yield criterion that satisfies the anisotropic presentation under uniaxial and equibiaxial tension [3]. Masse et al. [4] concluded that taking plastic anisotropy into account obviously improves the estimation of the final width. Many reports on the measurement of mechanical properties at the macro scale have been published. However, the in situ acquisition of the micro-scale mechanical properties of AZ31 alloys has yet to be conducted. Basu et al. [5] investigated size-dependent plastic responses by nanomechanical testing, high-resolution microscopy, and phase analysis and showed that the local microstructure has an obvious influence on small-scale plastic responses. Knowledge of the micro-scale mechanical properties of AZ31 alloys could provide a fundamental basis for its applications.

Most traditional experimental approaches, such as compression, tension, and torsion, are based on conventional bulk-scale testing, are destructive, and require a large number and volume of materials. The nanoindentation testing technique, a non- or semi-destructive approach, is considered a reliable, convenient, and robust approach with which to study the mechanical properties of materials at the nano- and micro-scales. Many scholars have proposed a series of simple mechanical properties for AZ31 alloys determined from indentation tests. However, in-depth studies of the mechanical properties of AZ31 alloys obtained via nanoindentation must be carried out, which can solve the impact due to the local deformation of the indentation test and complex operating conditions. In this study, nanoindentation tests were conducted using the continuous stiffness measurement technique (CSM), and the hardness of AZ31 alloys independent of size effects is obtained from the Nix model. The hardness values obtained are converted into strength values by using Tabor's factor, and the material parameters in the criteria were calculated to establish the anisotropic yield criterion of AZ31 alloys.

2. Materials and Experimental Procedure

The chemical composition of the material was Mg-90 wt.% Al-0.3 wt.% Zn-0.1 wt.%. The magnesium sheet was produced by the traditional rolling method. As shown in Figure 1, the specimens were machined into dimensions of 5 mm × 5 mm × 5 mm in the rolling direction (0°), diagonal direction (45°), transverse direction (90°), and vertical direction (ND). The nanoindentation tests were performed using a Nanoindenter G200 test system produced by Agilent Technologies with a triangular pyramid Berkovich diamond indenter. The optical microscopic images were performed using the GX53 Metallographic Microscope produced by OLYMPUS. Prior to indentation testing, the test surface was ground using a series of SiC sand papers with gradually finer grains and polished to a scratch-free mirror-like finish. In this study, the maximum indentation depth was set to 2000 nm and the indentation strain rate was 0.001. Each test was performed at room temperature and repeated thrice. The mean hardness was used for the following discussion. The maximum indentation depth and Poisson's ratio of the AZ31 alloys were set to 2000 nm and 0.28, respectively. A constant indentation strain rate was implemented by maintaining a constant loading rate ($\dot{p}/p$) during the test. The hardness values were obtained during loading in CSM mode [6,7], and the dwell time at a constant load of 45 mN was set to 100 s.

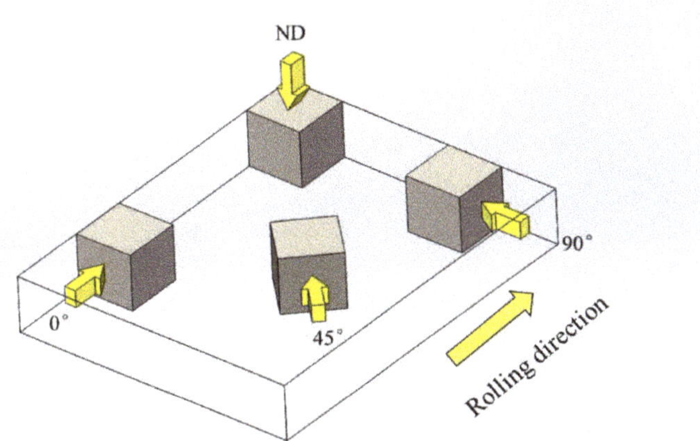

Figure 1. Schematic of the loading of rolled AZ31 alloys in four directions.

3. Results and Discussion

Figure 2 shows the optical microscopic images of the rolled AZ31 samples in the 0°, 45°, 90°, and ND directions (these directions are marked by the arrows in Figure 1); coarse grains of 5–20 µm could be observed in the samples. Barnett et al. found that grain size affects the mechanical responses of wrought magnesium [8]. The grain sizes of the rolled samples in the four directions were significantly different, which could explain the observed differences in their strength. The nanoindentation experiments were carried out in four directions, as shown in Figure 1.

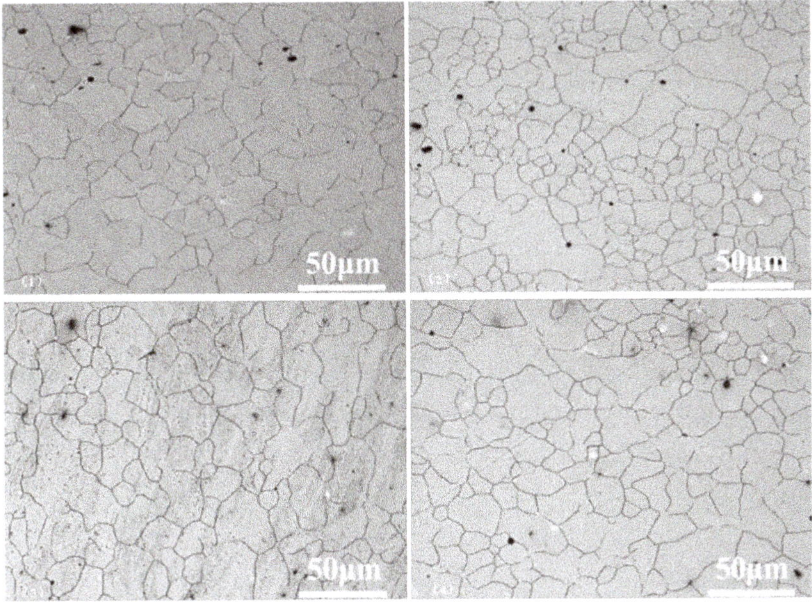

Figure 2. Optical microscopic images of the rolled AZ31 samples in the 0°, 45°, 90°, and vertical (ND) directions.

In CSM mode, the function of contact stiffness can be expressed as follows [6]:

$$S = \left[\frac{1}{\left(F_{amp}/h_{amp} \right) \cos \phi - \left(K_s - m\omega^2 \right)} - \frac{1}{K_f} \right]^{-1} \tag{1}$$

where S, F_{amp}, and h_{amp} are the contact stiffness, amplitude of the harmonic excitation force, and response displacement amplitude, respectively. In addition, ϕ is the phase shift, $\omega = 2\pi f$ is the angular frequency ($f = 45\text{Hz}$), and K_s, K_f, and m are the spring constant in the vertical direction, frame stiffness, and mass of the indenter, respectively. The contact stiffness increases nearly linearly with indentation depth in the three directions, as shown in Figure 3.

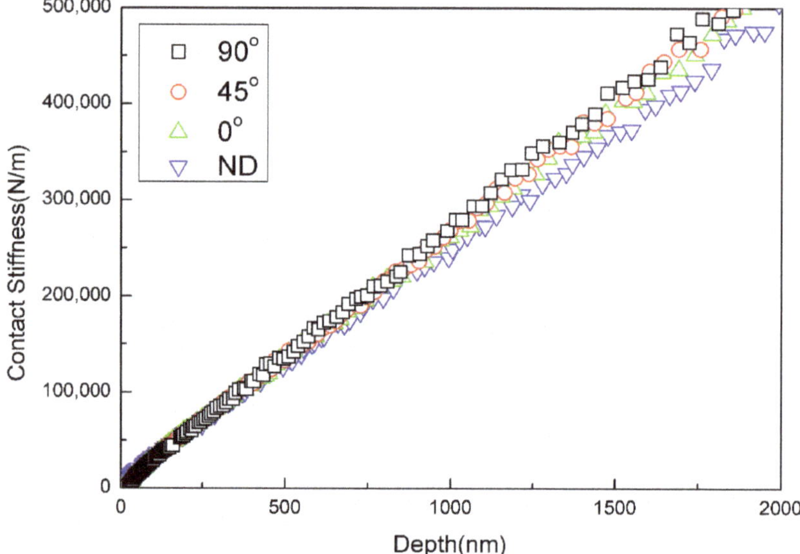

Figure 3. Nanoindentation contact stiffness-depth curves obtained during the loading of rolled AZ31 alloys in four directions.

The projected contact area A_c of a perfect Berkovich diamond indenter can be calculated as follows [9,10]:

$$A_c = 24.56 h_c^2 \tag{2}$$

where h_c is the contact depth and calculated by the equation:

$$h_c = h - \varepsilon \frac{P}{S} \tag{3}$$

where P and $\varepsilon = 0.75$ are the contact load and a contact for the Berkovich indenter, respectively. Hardness (H) is defined by the follow equation:

$$H = \frac{P}{A_c} \tag{4}$$

where P is the contact load. The hardness–depth curves of the samples in four directions can then be obtained. As shown in Figure 4, the experimental data of 45° have better repetition.

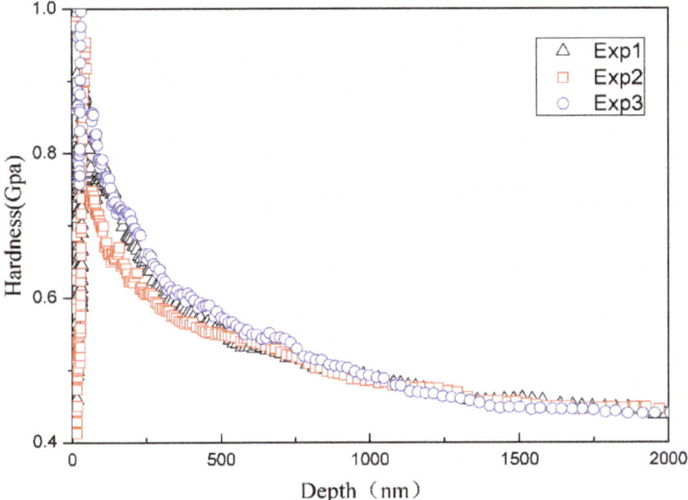

Figure 4. Nanoindentation hardness-depth curves of 45° from three repeated experiments.

The hardness values of rolled AZ31 alloys during loading in the 0°, 45°, 90°, and ND directions clearly differ, as shown in Figure 5. As the indentation depth increases, the hardness values decrease. The hardness increases as the angle increases from 0° to 90° and is smallest in the ND direction. Hardness is affected by the plate texture during rolling and shows significant anisotropy. Some factors of uncertainty and error in nanoindentation tests include surface roughness and surface texture, among others. In this study, the maximum penetration depth of the sample was approximately 2000 nm; such a depth is sufficiently large to assume that the contribution of surface inconsistencies is negligible. The hardness values observed in the 0°, 45°, 90°, and ND directions are listed in Table 1.

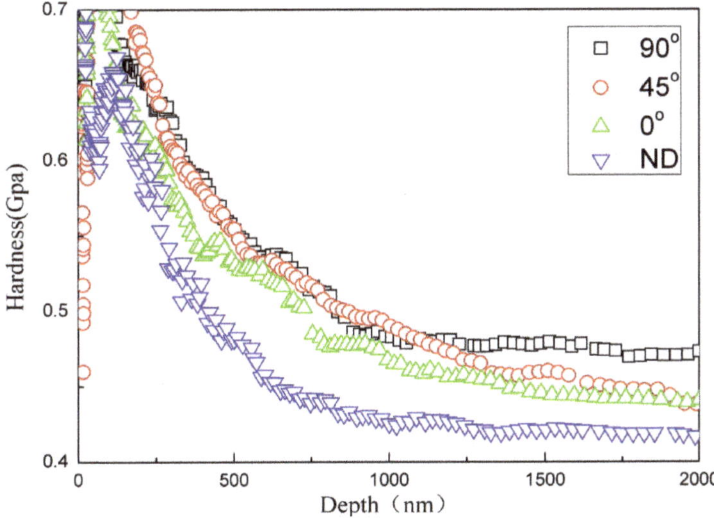

Figure 5. Nanoindentation hardness-depth curves obtained during the loading of rolled AZ31 alloys in four directions.

Table 1. Fitting values of hardness under different angles.

Angle (°)	0°	45°	90°	ND
H (GPa)	0.45	0.46	0.48	0.42

The indentation size effect refers to the variation in the indentation hardness (or indentation stress) as a function of indentation depth [11]. The effects of indentation depth on the mechanical properties of metals must be considered. Under the limit of infinite depth, Nix et al. [12,13] proposed a hardness model to investigate the indentation size effect of metal hardness as follows:

$$\frac{H}{H_0} = \sqrt{1 + \frac{h^*}{h}} \tag{5}$$

$$H^2 = H_0^2 h^* \cdot \frac{1}{h} + H_0^2 \tag{6}$$

where H_0 is the hardness in the limit of infinite depth and h^* is the characteristic length. The relationship between the square of hardness (H^2) and the reciprocal of indentation depth ($1/h$) should be linear, as shown in Equation (6). We performed the linear fitting of H^2 vs. $1/h$. Based on the indentation data collected between 500 and 1500 nm, the slope and intercept of the fitted straight line are $H_0^2 \cdot h*$ and H_0^2, respectively. Thus, H_0 independent of the indentation size effect can be obtained (Figure 6).

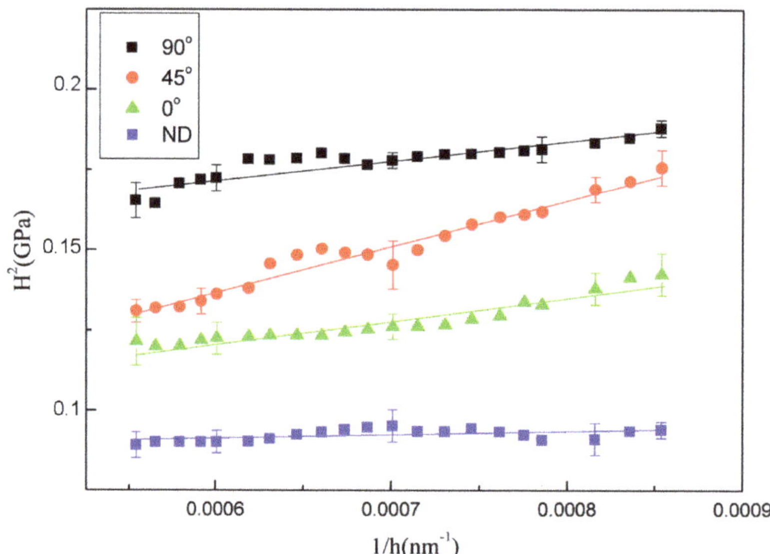

Figure 6. H^2 versus $1/h$ curves of the rolled AZ31 alloys in four angles.

H_0 may be determined from the results of linear fitting, as shown in Table 2.

Table 2. Fitting values of H_0 under different angles.

Angle (°)	0°	45°	90°	ND
H_0 (GPa)	0.35	0.36	0.42	0.28

A Tabor's factor of 3 can be used to convert the hardness values into yield stress values [14]:

$$H = 3\sigma \tag{7}$$

The strength of magnesium alloys can be obtained using Equation (7). The yield stresses in different directions are presented in Table 3

Table 3. Strength of the magnesium AZ31 alloys.

Angle (°)	0°	45°	90°	ND
σ_y (MPa)	116	120	137	94

The anisotropy yield criterion of rolled AZ31 magnesium alloys is established according to the Hill 1948 yield function. The three-dimensional yield function proposed by Hill [15] is defined as follows:

$$f_y = F_y(\sigma_{22} - \sigma_{33})^2 + G_y(\sigma_{33} - \sigma_{11})^2 + H_y(\sigma_{11} - \sigma_{22})^2 + 2L_y\sigma_{23}{}^2 + 2M_y\sigma_{31}{}^2 + 2N_y\sigma_{12}{}^2 - \bar{\sigma}_y{}^2 \tag{8}$$

where $F_y, G_y, H_y, L_y, M_y, N_y$ are used to describe direction-dependent yield stresses.

These coefficients can be obtained by considering the uniaxial compression at some angle relative to the rolling direction and denoting the uniaxial compressive yield stress σ_θ. The stress components in the Cartesian axis system are:

$$\sigma_{11} = \sigma_\theta \cos^2 \theta \quad \sigma_{12} = \sigma_\theta \sin \theta \cos \theta \quad \sigma_{22} = \sigma_\theta \sin^2 \theta \tag{9}$$

Substituting (9) for $\sigma_{11}, \sigma_{12}, \sigma_{22}$ in Equation (8) yields:

$$\sigma_\theta = ((1)((F_y + H_y) \sin^4 \theta) + (G_y + H_y) \cos^4 \theta - 2H_y \sin^2 \theta \cos^2 \theta + 2N_y \sin^2 \theta \cos^2 \theta)^{-1})^{1/2} \bar{\sigma}_y \tag{10}$$

Based on the primitive function, the uniaxial compressive yield stresses for the rolling direction (0°), diagonal direction (45°), and transverse direction (90°) are formulated as follows:

$$\begin{aligned}
\sigma_0 &= \left(\frac{1}{G_y + H_y}\right)^{1/2} \bar{\sigma}_y, \\
\sigma_{45} &= \left(\frac{4}{F_y + G_y + 2N_y}\right)^{1/2} \bar{\sigma}_y, \\
\sigma_{90} &= \left(\frac{1}{F_y + H_y}\right)^{1/2} \bar{\sigma}_y.
\end{aligned} \tag{11}$$

The yield stress in the rolled-plane direction, which is denoted σ_Z, can be concisely expressed as follows (12):

$$\sigma_z = \left(\frac{1}{F_y + G_y}\right)^{1/2} \bar{\sigma}_y \tag{12}$$

The following expression can be acquired by finding the primitive function, and four anisotropic parameters are formulated by solving Equations (11) and (12) as follows:

$$F_y = \frac{1}{2}\left(\frac{1}{\sigma_{90}^2} - \frac{1}{\sigma_0^2} + \frac{1}{\sigma_z^2}\right)\bar{\sigma}_y{}^2 \tag{13}$$

$$G_y = \frac{1}{2}\left(\frac{1}{\sigma_0^2} - \frac{1}{\sigma_{90}^2} + \frac{1}{\sigma_z^2}\right)\bar{\sigma}_y{}^2 \tag{14}$$

$$H_y = \frac{1}{2}\left(\frac{1}{\sigma_0^2} + \frac{1}{\sigma_{90}^2} - \frac{1}{\sigma_z^2}\right)\bar{\sigma}_y{}^2 \tag{15}$$

$$N_y = (\frac{2}{\sigma_{45}^2} - \frac{1}{2\sigma_z^2})\bar{\sigma}_y{}^2 \tag{16}$$

Through-thickness anisotropic parameters related to shear L_y and M_y are assumed to be identical to the anisotropic parameters of N_y and calculated as follows [16]:

$$L_y = M_y = N_y \tag{17}$$

The yield stress clearly increases as the angle along the rolling direction increases. This result is consistent with the results of macroscopic compression yield strength. The anisotropic parameters in the yield function can be calibrated using the data in Table 2. The yield stress in the vertical direction σ_z is used to represent the effective yield stress $\bar{\sigma}_y$. The calibrated anisotropic parameters are summarized in Table 4.

Table 4. Anisotropic parameters in the Hill yield function for rolled AZ31 alloys.

F_y	G_y	H_y	L_y	M_y	N_y
0.406	1.184	0.468	0.723	0.723	0.723

The anisotropic yield function determined via the nanoindentation test is utilized to describe the yield behavior of rolled AZ31 magnesium alloys. The following expression can be acquired by obtaining the primitive function:

$$\begin{aligned} f_y = {} & 0.406(\sigma_{22} - \sigma_{33})^2 + 1.184(\sigma_{33} - \sigma_{11})^2 + 0.468(\sigma_{11} - \sigma_{22})^2 + 0.723\sigma_{23}{}^2 \\ & + 0.723\sigma_{31}{}^2 + 0.723\sigma_{12}{}^2 - \bar{\sigma}_y{}^2 \end{aligned} \tag{18}$$

4. Conclusions

The Hill48 yield function is calibrated using data obtained from nanoindentation tests because the proposed plasticity model is applied to the rolling process, during which materials mainly experience compression. The mechanical behavior of the AZ31 alloys is investigated via nanoindentation tests with the CSM technique. The indentation hardness exhibits anisotropic behavior and is relatively large under high angles along the rolling direction, that is, the hardness values increase as the angle increases from 0° to 90°. The hardness in the ND direction is smallest among the values obtained. The anisotropic mechanical behavior of the alloys is analyzed, and the hardness independent of the effect of indentation depth is calculated. Tabor's factor can be used to convert hardness values into yield stress values. The Hill48 yield function of rolled AZ31 magnesium alloys is calibrated using the experimental results.

Author Contributions: Conceptualization, Z.W. and T.J.; methodology, T.J.; software, X.H.; validation, J.Q., X.L. and T.J.; formal analysis, X.S.; investigation, Z.W.; resources, Z.W. and T.J.; data curation, J.Q.; writing—originaldraft preparation, X.S.; writing—review and editing, T.J., Z.W. and X.H.; supervision, X.S.; project administration, X.L. and X.H.; funding acquisition, T.J. and X.H. All authors have read and agreed to the published version of the manuscript.

Funding: This research was funded by National Natural Science Foundation of China [11772215, 11772217, and 11802199]. Tao Jin is grateful to the Open Fund of the State Key Laboratory for Strength and Vibration of Mechanical Structures, Xi'An Jiaotong University [SV2019-KF-15] for its support. All financial contributions are gratefully acknowledged. And The APC was funded by [11772215].

Conflicts of Interest: Authors have no conflict of interest to declare.

References

1. Del Valle, J.; Pérez-Prado, M.; Ruano, O. Texture evolution during large-strain hot rolling of the Mg AZ61 alloy. *Mater. Sci. Eng. A* **2003**, *355*, 68–78. [CrossRef]
2. Haghshenas, M.; Wang, Y.; Cheng, Y.-T.; Gupta, M. Indentation-based rate-dependent plastic deformation of polycrystalline pure magnesium. *Mater. Sci. Eng. A* **2018**, *716*, 63–71. [CrossRef]
3. Hu, W. Characterized behaviors and corresponding yield criterion of anisotropic sheet metals. *Mater. Sci. Eng. A* **2003**, *345*, 139–144. [CrossRef]
4. Masse, T.; Chastel, Y.; Montmitonnet, P.; Bobadilla, C.; Persem, N.; Foissey, S. Impact of mechanical anisotropy on the geometry of flat-rolled fully pearlitic steel wires. *J. Mater. Process. Technol.* **2011**, *211*, 103–112. [CrossRef]
5. Basu, I.; Ocelík, V.; De Hosson, J.T.M. Size dependent plasticity and damage response in multiphase body centered cubic high entropy alloys. *Acta Mater.* **2018**, *150*, 104–116. [CrossRef]
6. Fischer-Cripps, A.; Johnsonm, K. Introduction to Contact Mechanics. Mechanical Engineering Series. *Appl. Mech. Rev.* **2002**, *55*, 3–13. [CrossRef]
7. Xiao, G.; Liu, E.; Jin, T.; Shu, X.; Wang, Z.; Yuan, G.; Yang, X. Mechanical properties of cured isotropic conductive adhesive (ICA) under hygrothermal aging investigated by micro-indentation. *Int. J. Solids Struct.* **2017**, *122*, 81–90. [CrossRef]
8. Barnett, M.R.; Keshavarz, Z.; Beer, A.G. Influence of grain size on the compressive deformation of wrought Mg–3Al–1Zn. *Acta Materialia* **2004**, *52*, 5093–5103. [CrossRef]
9. Xiao, G.S.; Yuan, G.Z.; Jia, C.N.; Yang, X.X.; Li, Z.G.; Shu, F.X. Strain rate sensitivity of Sn-3.0Ag-0.5Cusolder investigated by nanoindentation. *Mater. Sci. Eng. A* **2014**, *613*, 336. [CrossRef]
10. Qiu, J.; Xiao, G.; Jin, T.; Su, B.; Shu, X.; Ma, S. Indentation Strain Rate Sensitivity of CoCrFeNiAl0.3 High-Entropy Alloy. *Adv. Eng. Mater.* **2019**, *21*, 21. [CrossRef]
11. Vemuri V, K. Springer Science+Business Media. *J. Inf. Technol. Cases Appl.* **2019**, *21*, 109–112.
12. Nix, W.D.; Gao, H. Indentation size effects in crystalline materials: A law for strain gradient plasticity. *J. Mech. Phys. Solids* **1998**, *46*, 411–425. [CrossRef]
13. Xiao, G.; Yang, X.; Qiu, J.; Chang, C.; Liu, E.; Duan, Q.; Shu, X.; Wang, Z.-H. Determination of power hardening elastoplastic constitutive relation of metals through indentation tests with plural indenters. *Mech. Mater.* **2019**, *138*, 103173. [CrossRef]
14. Tabor, D. A simple theory of static and dynamic hardness. In *Proceedings of the Royal Society of London*; Series A, Containing Papers of a Mathematical and Physical Character; The Royal Society: London, UK, 1948; Volume 192, pp. 247–274.
15. Hill, R. A theory of the yielding and plastic flow of anisotropic metals. In *Proceedings of the Royal Society of London*; Series A, Containing Papers of a Mathematical and Physical Character; The Royal Society: London, UK, 1948; Volume 193, pp. 281–297.
16. Wang, G.; Qian, X.; Li, X.; Hou, H.; Liu, Y.; Lou, Y. A Study on Compressive Anisotropy and Nonassociated Flow Plasticity of the AZ31 Magnesium Alloy in Hot Rolling. *Math. Probl. Eng.* **2014**, *2014*, 1–9. [CrossRef]

Publisher's Note: MDPI stays neutral with regard to jurisdictional claims in published maps and institutional affiliations.

Article

Degradation of a Micro-Hybrid Dental Composite Reinforced with Polyaramide Fiber under the Influence of Cyclic Loads

Leszek Szalewski [1,†], Aneta Kamińska [2,†], Eliza Wallner [3], Justyna Batkowska [4], Tomasz Warda [5], Dorota Wójcik [6,*] and Janusz Borowicz [6]

[1] Chair of Integrated Dentistry, Medical University of Lublin, 20-093 Lublin, Poland; leszek.szalewski@umlub.pl

[2] Independent Laboratory of Dental Techniques, 20-093 Lublin, Poland; aneta.kaminska@umlub.pl

[3] Chair and Department of Jaw Orthopedics, Medical University of Lublin, 20-093 Lublin, Poland; eliza.wallner@umlub.pl

[4] Institute of Biological Basis of Animal Production, University of Life Sciences in Lublin, 20-950 Lublin, Poland; justyna.batkowska@up.lublin.pl

[5] Faculty of Mechanical Engineering, Lublin University of Technology city, 20-618 Lublin, Poland; t.warda@pollub.pl

[6] Department of Dental Prosthetics, Medical University of Lublin, 20-093 Lublin, Poland; janusz.borowicz@umlub.pl

[*] Correspondence: dorota.wojcik@umlub.pl

[†] Leszek Szalewski, Aneta Kamińska—these authors contributed equally to this work.

Received: 8 September 2020; Accepted: 12 October 2020; Published: 19 October 2020

Featured Application: The article can help dentists in deciding whether to choose fiber-reinforced composite adhesive bridges as a treatment option. The results indicate that the choice of another therapeutic method should be considered in the case of increased occlusal forces (parafunction, bruxism, occlusal obstructions). The obtained results may become a starting point for research on new materials/fibers to increase the mechanical properties of adhesive composite bridges to obtain a material resistant to high occlusal forces.

Abstract: Dental composites reinforced with glass fibers have a low tensile modulus and relatively low fatigue resistance. The aim of the study was to analyze the fatigue properties of a dental composite reinforced with polyaramide fibers under the influence of a cyclic, vertical load. For this purpose, we designed a thermoformable template, corresponding to the construction of adhesive bridges in the side section of the jaw. Fifty-four composite samples were made for the study. They were divided into three groups—control (K) and two experimental groups (R1 and R2). The experimental samples were subjected to cyclic fatigue using 75 N load. The number of cycles was 4690 and 20,100. The study used a three-point bending test. Statistical analysis showed a change in elasticity in groups related to the number of load cycles. The study showed that the samples from the control group required the greatest force to break in relation to those subjected to the work cycles. The maximum force in control (K) group was 738.1 N, R1—487.8 N, and R2—451.4 N. The determined algorithm showed a change in deflection associated with the increase of force value. The study did not show any relationship between the type of sample fracture and the number of load cycles.

Keywords: composite resins; compressive strength; fixed partial denture

1. Introduction

Increasing aesthetic requirements of patients has led to the development of composite dental resins. Usually dental composites are made up from the matrix and fillers that are connected with each other by so-called silanes. Modifications of these two components in the last 20 years have increased the use of dental composites. They are a material commonly used to reconstruct lost tooth tissues using aesthetic restorations [1,2].

Composite materials are based on methacrylate compounds. Their matrix is an organic photopolymerizating resin, which consists mostly of bisphenol A-glycidyl methacrylate (Bis-GMA), triethylene glycol dimethacrylate (TEGMA), and urethane dimethacrylate (UDMA). The inorganic phase is macro-, micro-, or nanofillers, based mostly on silicon compounds [3,4]. Additionally, the composite contains photo initiators and proadhesive agents—silanes. The micro-hybrid composites contain a mixture of at least two types of "glass" or quartz molecules, irregular in shape and similar in diameter (from 0.2 to 3 μm) and from 5 to 15% of small particles (0.04 μm). In these materials, the filler constitutes 60–70% by volume, i.e., about 77–84% by weight of the composite part [5].

In order to increase the field of application, as well as to broaden the indications for the use of composite materials, their combinations with fibers such as carbon, polyaramide, polyethylene, and glass are used [6,7]. Currently, such solutions are used in periodontology [8], endodontics [9], prosthetics [10], and orthodontics [11].

Both wholly aromatic polyamides or the shorter aromatic polyamids form aramids and stand for synthetic polyamides comprising >85% amide groups (–CO–NH–) bound directly to two aromatic rings. Such polymers are characterized as high-performance materials due to their very good mechanical strength and exceptional high thermal resistance. They are spun into fibers and used in advanced fabrics, such as sport and work protective clothing, bullet-proof armor, advanced composites in armament and aerospace industries, composites such as asbestos substitutes, and high-temperature insulation paper.

The aramid structure is based on rigid aromatic amide linkage. It is responsible for the exceptional properties of these materials. Highly directional and efficient interchain hydrogen bonds are established, giving basis to materials with a high tendency to crystallize and with extremely high cohesive energy density. At the same time, they are also responsible for the insolubility of the wholly aromatic polyamides, a disadvantage that prevents the expansion of the application field of these materials, whereas the improvement of the solubility is a topic of the present research interest [12]. Aromatic polyamides are obtained in reactions that form amide bonds between aromatic rings of high thermal stability and high strength. The outstanding rigid molecular chain structure, good orientation, and organization of the crystalline structure provides high strength and low elongation of the orientation of the molecular chains. This provides high tensile strength, impact, and differentiated thermal stability for various temperature ranges for an extended time [13]. The covalent bonds in the polymer are responsible for the high strength. However, man-made polymers generally do not exhibit the corresponding potential high modulus. High modulus and strength may result from structural perfection, crystallinity, and crystalline and amorphous orientation. It is well known that the highest elastic moduli reported from linear polymers are generally much smaller than theoretical values [14].

The bending strength of a prosthetic restoration is an extremely important feature whose assessment is necessary in specific cases of prosthesis use, for example, when it is necessity to increase its resistance to fracture during a long period of use or when a prosthesis with an elongated structure is made [15]. Chewing is the main factor that causes mechanical degradation of the composite resin. While chewing, the mandible moves repeatedly in the vertical and horizontal planes [16,17]. The mechanics of the chewing process adversely affect composite resins and lead to stress within the material structure [18]. During chewing, prosthetic restorations are subjected to loads with forces whose direction is consistent with the long axis of the tooth. The size of chewing force in the mouth is from 3 to 36—49 N on average—but it can also reach 1000 N [19,20]. During this process, the opposing teeth remain in mutual point contact, and the highest loads occur in the mouth only sporadically [21].

The number of cyclic contacts during chewing and swallowing per day depend on the number of meals, their consistency, the person's age and sex, the number of natural teeth, the type of prosthetic restorations, and the functional disorders of the masticatory apparatus that may significantly increase the number of contacts [22,23]. On average, single contacts last 0.3 s, and their number is variable, ranging from 658 to 2300 contacts a day [21,24,25].

The available literature has not yet explained all aspects of the degradation of composites nor, consequently, the problem related to the maximum time of the restoration use in real conditions. In general, tests have been carried out on standardized samples, reproducing cyclic loads and finite elements that were usually subjected to static loads [26–29].

The aim of the research was to determine and compare the effect of cyclic vertical loads on the strength of composite resins reinforced with high molecular weight polyaramide fiber.

2. Materials and Methods

2.1. Preparation of the Test

The study used a gypsum model of the jaw, from which a fragment covering the area of three teeth was dissected, between the second premolar and the second molar on the left side. The obtained model (model A) was duplicated in a silicone mass, and the obtained form was filled with a chemo-hardening, burnt-out acrylic resin, which was changed into a chromium–nickel alloy in the casting process. On the basis of the obtained model, we produced standard templates (Figure 1) that were used to make samples. Plates that were 1.5 mm thick were used to make the templates, which were placed in a pressure thermoforming device. The tiles were heated for 55 s, after which they were stretched on the prepared models. Next, in the gypsum model A, the first molar tooth was removed, and in the remaining two teeth, limiting this lack, a preparation was made within the crowns to obtain a model that was reproduced and turned into a chromium–nickel alloy (model B). Model B was used as a handle in which the samples were fixed during the test. In order to obtain the model on which the samples were made (C-model), before replacing with the chromium–nickel alloy, we blocked a 2-mm space for the future restored tooth (bridge span) with the use of a chemically curing resin. The obtained distance prevented the sample from leaning against the model during its possible deflection during the test (Figure 1). A chromium–nickel model (D) (Figure 1) of the opposed tooth (the first left mandible molar) was also made for model A, which allowed us to keep the proper contact of the opposing teeth. This model was used in the study as a contact element with the test samples.

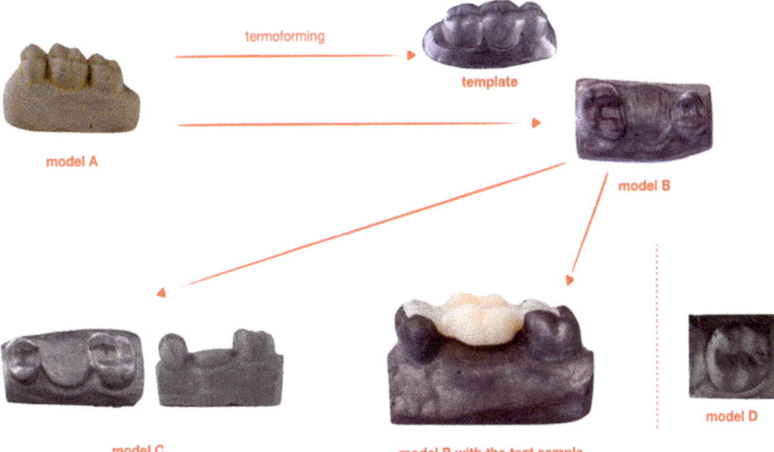

Figure 1. The schema of sample preparation.

The study used a technical composite in A2 color and a polyaramide fiber with a width of 3 mm, which were used to made adhesive bridges.

2.2. Specimen Preparation

On model C, we pressed thin thermoformable films to insulate the material from the model walls. Then, the first layer of 1 mm thick composite was applied to the prepared place, on which a 17 mm polyaramide fiber was placed, corresponding to the range between the prepared areas in crowns of the teeth defining the gap (sample length). The template was then applied and polymerized using a 470 nm wavelength tube (Clear Blue LED 1200 mW/cm^2) for 40 s for every 5 mm section. Subsequent layers of 0.5 mm thick composite were applied and condensed using standard dental applicators and polymerized in the same way until the template was filled. The excess material was removed. After polymerization and obtaining anatomical shapes, we released the samples from the models, and any excess of the composite material was removed using a composite milling cutter. The value of the width and height of connectors in the samples was 4 × 3 mm, respectively.

The research material consisted of anatomical samples (n = 54), which were divided into three groups: the control group (K, 21 samples) and two experimental groups (R1, R2), divided according to the number of cycles: 18 and 15 samples, respectively.

2.3. Fatigue Strength Test

The Zwick/Roell Z 2.5 (ZwickRoell GmBh & Co. KG, Ulm, Germany) testing machine was used in the test. During strength tests, we placed all samples in the holder, which was model C, with a distance of 11.5 mm between two end supports. When testing the control samples, we set the initial force to 5 N, and the traverse speed was 5 mm/min. The test lasted until the sample was destroyed, which was defined as a 20% drop in strength in relation to the achieved maximum force for a given sample.

Samples from R1 and R2 were subjected to cyclic loads of 75 N at a 90° angle using the tooth's opposed model. During this process, the traverse speed was 10 mm/min, and the sample holding time was 0.3 s. After this time, the tested object was relieved to zero force. Then, the cycle was repeated. The number of cycles in the R1 group was 4690, which is the average value corresponding to the number of weekly opposing teeth contact, and in the R2 group, we used 20,100 cycles, the number of which corresponds to monthly dental contacts [21]. After fatigue loads, the samples were subjected to a bending strength test in the same way as in group K. The obtained parameters were maximum force [N], deflection at maximum force [mm], deflection at yielding point [mm], and post-yield displacement [mm].

2.4. Statistical Analysis

The data were analyzed using the SPSS 20.0 PL statistical package (IBM, New York, NY, USA) [30], the Kolmogorov–Smirnov test (normality of distribution), and a one-way analysis of variance were used with Tukey's multiple comparison test. Spearman's nonparametric correlation coefficients and selected linear regression coefficients were also estimated. The distribution of the crack types recorded in individual samples was verified by a non-parametric χ^2 test.

2.5. Microscope Observations

After durability tests, microscopic observations of the work surfaces of fractures of samples were made to check for possible damage. Observations were made using a VEGA//LMU scanning electron microscope with 100 and 2000 magnification.

3. Results

The study showed that the samples from the control group required the greatest force to break in relation to those subjected to the work cycles. The study did not show statistically significant differences between the destructive force in the case of groups R1 and R2, despite the differences in the number of load cycles; however, the numerical differences were found in favor of group treated by smaller number of cycles. Moreover, tests from experimental groups (R1 and R2) were characterized by significantly lower elasticity (deflection) under the maximum force action. Samples from experimental groups were definitely more inflexible compared to the control samples. The reduction of deflection after cyclic loads may result from changes in the internal structure of the samples. Detailed results regarding strength parameters for particular groups are presented in Table 1.

Table 1. Strength parameters of samples included in the tests (mean ± standard error of the mean (SEM)).

Parameter	K	R1	R2
Maximum force [N]	738.1 [a] ± 50.3	487.8 [b] ± 28.8	451.4 [b] ± 40.5
Deflection at maximum force [mm]	1.204 [a] ± 0.115	0.605 [b] ± 0.049	0.630 [b] ± 0.106
Deflection at yielding point [mm]	0.907 [a] ± 0.075	0.483 [b] ± 0.043	0.436 [b] ± 0.018
Post-yield displacement [mm]	1.210 [a] ± 0.112	0.568 [b] ± 0.040	0.614 [ab] ± 0.107

[a, b]—differ significantly at ≤0.05; SEM—standard error of the mean.

During the strength tests, we found three types of cracks: longitudinal, transversal, and defragmentation (Figure 2). It was not demonstrated that the method of sample fracture depended on the number of performed fatigue cycles (χ^2, $p = 0.132$). Cracks in external polyaramide fibers were observed only in one sample from group R2 (Figure 3). Photographs from the electron microscope show the fiber structure, which is characterized by an empty space inside. Effective forces are focused on the outer surface of the fiber. The presence of fine cracks in the area of the main crack was also noticed. The vertical application of force causes the emergence of cutting forces, leading to side cracks in various directions. It has been observed that with the increase in the number of fatigue cycles, the number of these cracks is much higher, which is related to the increase in the hardness of the composite material with the simultaneous increase in brittleness (Figures 4–6). Such a situation may result from the fact that the contact points occur on the unevenness of the teeth chewing surfaces and may be a consequence of the lack of perfect stiffness of the test sample system.

Table 2 presents correlation coefficients between the analyzed strength parameters. There was a statistically significant dependence ($p \leq 0.01$) between the number of cycles and individual measured characteristics, whose values decreased considerably with the increase of the number of destructive cycles. It has been noted that with the increase of the maximum force, the flexibility of the sample characterized by material deflection increased significantly. Similarly, highly significant, positive relationships were found for all three plastic deformations of the analyzed samples.

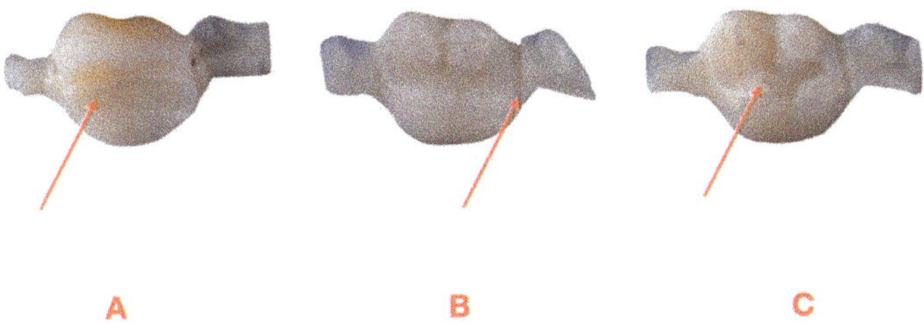

Figure 2. The types of ruptures found during the experiment ((**A**) longitudinal, (**B**) transverse, (**C**) defragmentation).

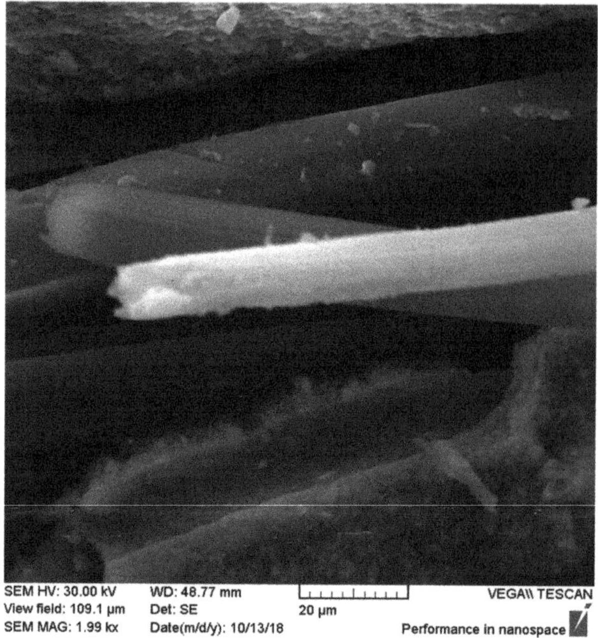

Figure 3. View of sample R2 (magnification 1.99 k×).

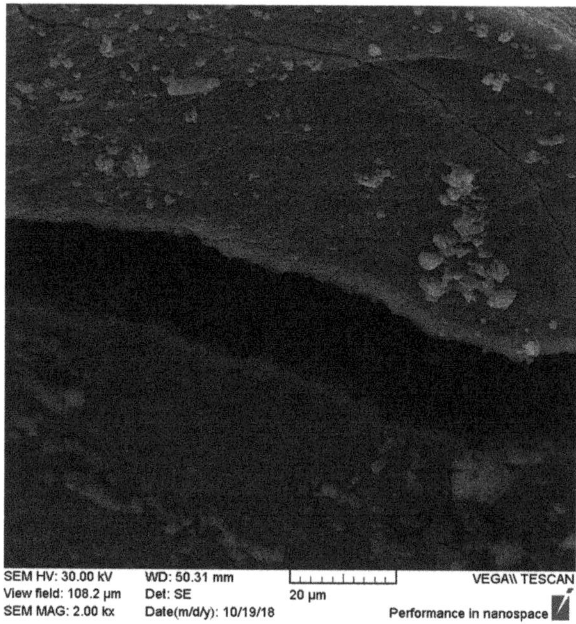

Figure 4. View of control group sample (magnification 2.00 k×).

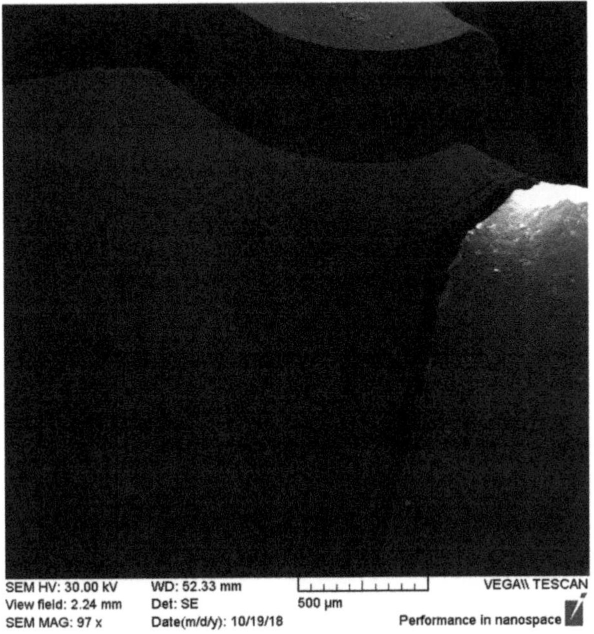

Figure 5. View of sample R1 (magnification 97×).

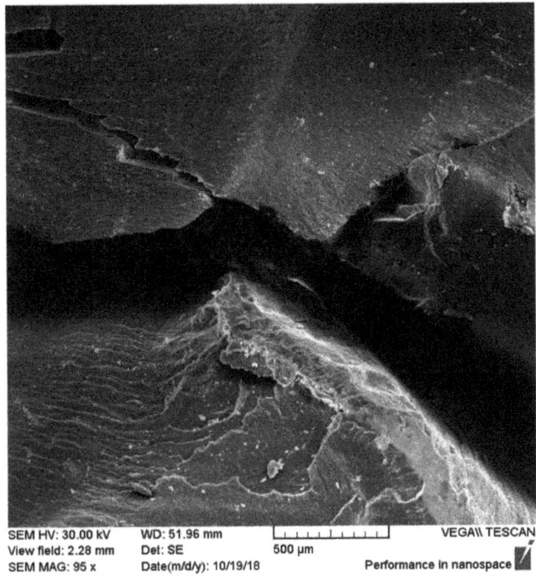

Figure 6. View of sample R2 (magnification 95×).

Table 2. The correlation coefficients between measured samples' characteristics.

	No. of Cycles	Maximum Force	Deflection at Maximum Force	Deflection at Yielding Point
Maximum force	−0.577 **			
Deflection at maximum force	−0.608 **	0.853 **		
Deflection at yielding point	−0.703 **	0.747 **	0.759 **	
Post-yield displacement	−0.666 **	0.822 **	0.979 **	0.780 **

** correlation is significant at $p \leq 0.01$ (one-sided).

Figure 7 illustrates the dependence of deflection at maximum force on the value of this force depending on the group treated by various numbers of working cycles. It is visible that each group demonstrated different reactions on breaking force. The most resistant was the control group not subjected to cyclic loads. The durability of other groups also varied according to number of cycles, with better resistance characterized by R1 group in comparison to the R2 group.

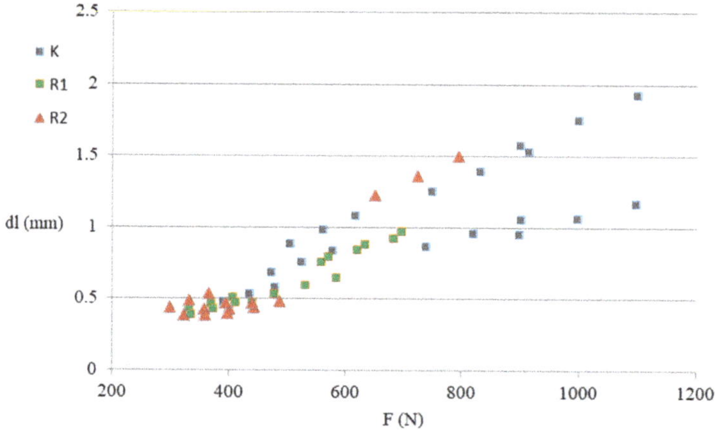

Figure 7. The relationship between deflection at maximum force and the maximum force value.

Figure 8 shows that the dependence of the deflection on the load value in groups K, R1, and R2 were distinguished on the basis of the fatigue cycle number. It is possible that materials in certain samples could behave the other way; however, we did not find any differences using standard error of the mean (SEM). All determined regression coefficients were statistically significant ($p \leq 0.01$). Their properties resulted in the possibility of estimating the expected value of a variable. Therefore, on this basis, expected values of deflection were estimated under the influence of a specific load depending on the number of cycles (Table 3). Increasing the value of the acting force may lead to an increase in sample deflection in each of the studied groups.

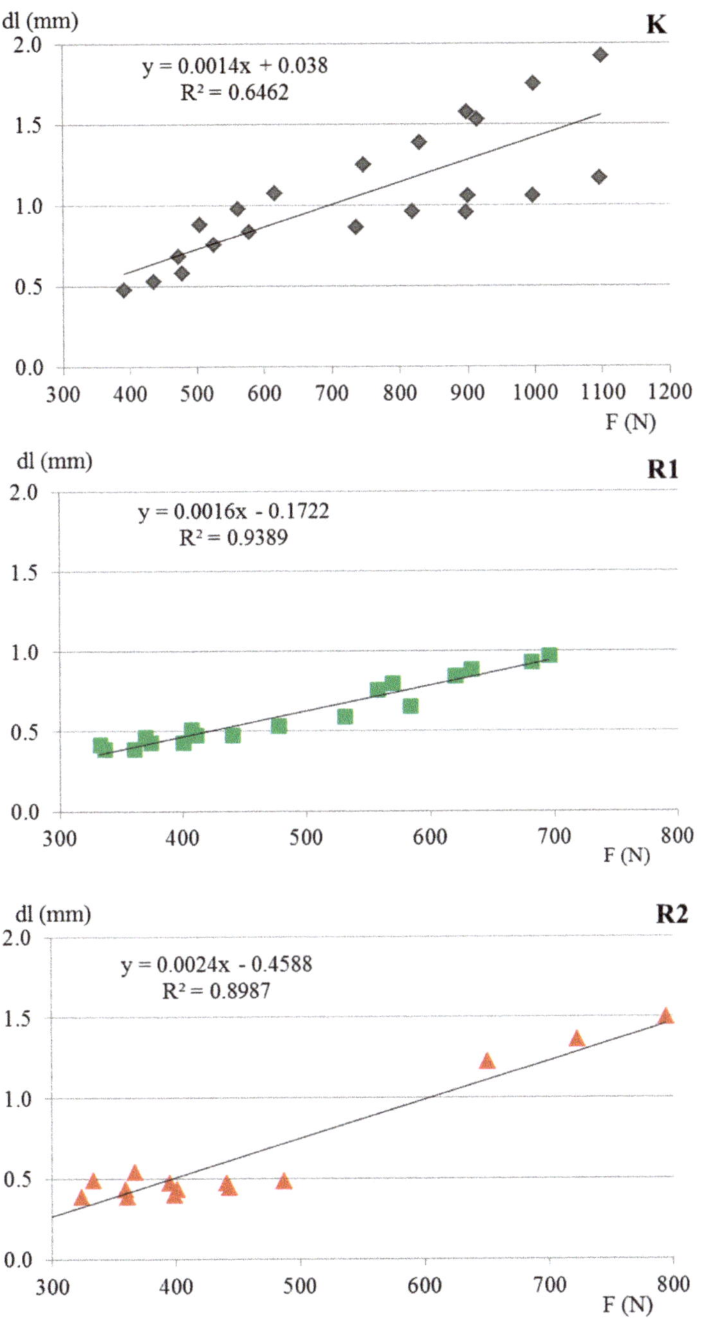

Figure 8. Relationship of deflection from the load values in particular groups: K, R1, R2.

Table 3. Expected values of deflection depending on the specific load (mm).

Group	F (N)			
	500	600	800	1000
K	0.659	0.799	1.079	1.359
R1	0.606	0.766	1.086	1.406
R2	0.76	1.01	1.51	2.01

4. Discussion

Fiber-reinforced composite materials are used as alternative materials for metal restorations. Modifications of composite resins, through various types of fibers improving the mechanical properties of restorations, allow for their use not only in prosthetic crowns, but also in solid partial dentures (FPD)—e.g., in adhesive bridges [31]. These additions differ from the total permanent dentures by the replacement of the crowns, on which the prosthetic bridge structure is based, with crown inserts, i.e., onlay and inlay. This construction of prosthetic bridges is an alternative for patients who do not agree to a complete permanent replacement due to the need of significant tooth tissue reduction, or when minimal tooth reduction is possible [32,33]. In the construction of bridges, the connectors between the individual elements of the bridge are most vulnerable to destruction. Solid partial dentures of fiber-reinforced composite resins are made by placing the fiber in the structure and surrounding it with a composite resin. During their fatigue cyclic tests, Lobhauer et al. [34] observed the slow spread of cracks in brittle materials, such as composite resins. The literature lacks studies describing polyaramid fiber due to the fact that this is a new material in dentistry. Research results derived from the study on composites reinforced with polyaramid fibers provided by Selvaraj et al. acknowledge that the addition of polyaramid into composite impact/increases composite strength [35].

Previous studies on dental composites used to make prosthetic bridges were mainly based on standardized samples. Meanwhile, such samples subjected to cyclic loads show lower values of bending strength (30–50%) than those obtained in static studies and are considered more sensitive in assessing the effectiveness of clinical materials [29]. When testing the strength of dental composites, Papadogiannis et al. [28] stated that fatigue strength is associated not only with the type of filler, but also with the resin matrix. Kuroda et al. [27] used standardized samples with dimensions of $3.0 \times 4.0 \times 40$ mm in their tests, which underwent cyclic fatigue loads with a force of 100 N. They noted that the strength of fiber-reinforced dental composites increases the bending strength. They noted better strength of composites reinforced with glass fiber under the influence of increasing force compared to the unreinforced control sample. However, these studies did not take into account the physiological interactions between the opposing teeth during chewing. In a study by Nobuhisa et al. [26], who analyzed the FEM model of a three-point conventional bridge reinforced with fiber glass, the authors applied a load of 629 N in the rebuilding the lack of the first molar in the mandible. The significant improvement in connector rigidity under vertical load conditions causing twisting and bending movement was stated. As a consequence, the stiffness of connections between individual elements of the sample structure improved, which significantly reduced the deflection of the span because the stresses generated by the vertical load were transferred to the reinforcing fiber. However, no relationship was found between the number of fatigue cycles and bending strength. Similar dependences on the better strength of samples reinforced with polyaramide fiber were recorded in the present research, whereby the actual strength of the manufactured bridges was tested with a strictly established number of loading cycles. Although the experimental groups required significantly lower force in order to break, the destructive force was still in the reported range of habitual and maximal biting forces in molar teeth, which are 300 N and 500 N, respectively [36,37].

The authors are aware that the tests were carried out in laboratory conditions and do not perfectly reflect the conditions in the oral cavity (the authors did not use any bonding system between sample and the chromium–nickel model); nevertheless, the demonstrated differentiation indicates the validity of further research in this direction.

5. Conclusions

Insertion of numerous and fine polyamide fiber fragments into the composite resin may contribute to the elimination of cracks in the material itself, limiting their propagation in restorations. Such a reinforcement could contribute to extending the life of these additions. The improvement of mechanical properties of dental composites, including flexural strength, is of particular importance in the rehabilitation of patients with functional disorders of the masticatory apparatus, where chewing forces are very high.

Studies based on the analysis of anatomically reversible elements require the introduction of laboratory validation to allow comparing test results. The conditions for dental materials in the oral cavity are difficult. In addition, the large anatomical variability of elements reconstructed with the use of dental materials excludes the possibility of developing a conclusion on the basis of standard solutions.

The deflection of adhesive bridges predicted in the study indicates the need for further research in this field with the use of finite element modelling. The polyaramide fiber used in the research can replace the previously used glass fiber to strengthen the construction of prosthetic bridges.

Author Contributions: L.S.: conceptualization, data curation, formal analysis, methodology, writing—original draft; A.K.: conceptualization, data curation, formal analysis, methodology, writing—original draft; E.W.: data curation, methodology, resources, writing—original draft; J.B. (Justyna Batkowska): data curation, formal analysis, visualization, writing—original draft; T.W.: conceptualization, data curation, methodology, visualization, writing—review and editing; D.W.: data curation, formal analysis, methodology, visualization, writing—original draft; J.B. (Janusz Borowicz): conceptualization, formal analysis, supervision, writing—review and editing. All authors have read and agreed to the published version of the manuscript.

Funding: This research received no external funding.

Conflicts of Interest: The authors declare no conflict of interest.

References

1. Cramer, N.B.; Stansbury, J.W.; Bowman, C.N. Recent advances and developments in composite dental restorative materials. *J. Dent. Res.* **2011**, *90*, 402–416. [CrossRef] [PubMed]
2. Khurshid, Z.; Zafar, M.; Qasim, S.; Shahab, S.; Naseem, M.; Abu Reqaiba, A. Advances in nanotechnology for restorative dentistry. *Materials* **2015**, *8*, 717–731. [CrossRef] [PubMed]
3. Wacławczyk, A.; Postek-Stefańska, L.; Pietraszewska, D.; Birkner, E.; Zalejska-Fiolka, J.; Wysoczańska-Jankowicz, I. TEGDMA and UDMA monomers released from composite dental material polymerized with diode and halogen lamps. *Adv. Clin. Exp. Med.* **2018**, *27*, 469–476. [CrossRef]
4. Manojlovic, D.; Radisic, M.; Vasiljevic, T.; Zivkovic, S.; Lausevic, M.; Miletic, V. Monomer elution from nanohybrid and or mocer-based composites cured with different light sources. *Dent. Mater.* **2011**, *27*, 371–378. [CrossRef] [PubMed]
5. LeSage, B.P. Aesthetic anterior composite restorations: A guide to direct placement. *Dent. Clin. North. Am.* **2007**, *51*, 359–378. [CrossRef] [PubMed]
6. Sfondrini, M.F.; Massironi, S.; Pieraccini, G.; Scribante, A.; Vallittu, P.K.; Lassila, L.V.; Gandini, P. Flexural strengths of conventional and nanofilled fiber-reinforced composites: A three-point bending test. *Dent. Traumatol.* **2014**, *30*, 32–35. [CrossRef] [PubMed]
7. Acharya, P.P.; Bhat, R. The physiochemical properties of dental resin composites reinforced with milled E-glass fibers. *Silicon* **2018**, *10*, 1999–2007.
8. Akcalı, A.; Gümüş, P.; Özcan, M. Clinical comparison of fiber-reinforced composite and stainless-steelwire for splinting periodontally treated mobile teeth. *Braz. Dent. Sci.* **2014**, *17*, 39–49. [CrossRef]
9. Albashaireh, Z.S.; Ghazal, M.; Kern, M. Effects of endodontic post surface treatment, dentin conditioning, and artificial aging on the retention of glass fiber-reinforced composite resin posts. *J. Prosthet. Dent.* **2010**, *103*, 31–39. [CrossRef]

10. Bijelic-Donova, J.; Garoush, S.; Lassila, L.V.J.; Keulemans, F.; Vallittu, P.K. Mechanical and structural characterization of discontinuous fiber-reinforced dental resin composite. *J. Dent.* **2016**, *52*, 70–78. [CrossRef]

11. Vallittu, P.K. An overview of development and status of fiber-reinforced composites as dental and medical biomaterials. *Acta Biomater. Odontol. Scand.* **2018**, *4*, 44–55. [CrossRef] [PubMed]

12. Reglero Ruiz, J.A.; Trigo-López, M.; García, F.C.; García, J.M. Functional aromatic polyamides. *Polymers* **2017**, *9*, 414. [CrossRef] [PubMed]

13. Konarzewski, V.H.C.; Spiekemann, F.L.; Santana, R.M.C. Natural ageing of polyaramide fiber from ballistic armor. *Polímeros* **2019**, *29*, 1–7. [CrossRef]

14. Vasanthan, N. Polyamide fiber formation: Structure, properties and characterization. In *Handbook of Textile Fibre Structure*; Woodhead Publishing: Sawston, UK, 2009; pp. 232–256.

15. Digholkar, S.; Madhav, V.N.V.; Palaskar, J. Evaluation of the flexural strength and microhardness of provisional crown and bridge materials fabricated by different methods. *J. Indian Prosthodont. Soc.* **2016**, *16*, 328–334.

16. Pieniak, D.; Niewczas, A. Phenomenological evaluation of fatigue cracking of dental restorations under conditions of cyclic mechanical loads. *Acta. Bioeng. Biomech.* **2012**, *14*, 9–17.

17. Hunicz, J.; Niewczas, A.; Kordos, P.; Pieniak, D. Experimental test stand for analysis of composite dental fillings degradation. *Eksploat. Niezawodn. Maint. Reliab.* **2007**, *37*, 37–43.

18. Sarveshkumar, J.; Jayalakshmi, S. Bond failure and its prevention in composite restoration—A review. *J. Pharm. Sci. Res.* **2016**, *8*, 627–631.

19. Ramalho, A.; Antunes, P.V. Reciprocating wear test of dental composites against human teeth and glass. *Wear* **2007**, *263*, 1095–1104. [CrossRef]

20. Koc, D.; Dogan, A.; Bek, B. Bite force and influential factors on bite force measurement a literature review. *Eur. J. Dent.* **2010**, *4*, 223–232. [CrossRef]

21. Chitmongkolsuk, S.; Heydecke, G.; Stappert, C.; Strub, J.R. Fracture strength of all-ceramic lithium disilicate and porcelain-fused-to-metal bridges for molar replacement after dynamic loading. *Eur. J. Prosthodont. Restor. Dent.* **2002**, *10*, 15–22.

22. Pereira, L.J.; Gaviao, M.B.D.; Bonjardim, L.R.; Castelo, P.M.; Van der Bilt, A. Muscle thickness, bite force, and craniofacial dimensions in adolescents with signs and symptoms of temporomandibular dysfunction. *Eur. J. Orthod.* **2007**, *29*, 72–78. [CrossRef]

23. Rosa, L.B.; Bataglion, C.; Siéssere, S.; Palinkas, M.; Mestriner, W., Jr.; De Freitas, O.; de Rossi, M.; de Oliveira, L.F.; Regalo, S.C. Bite force and masticatory efficiency in individuals with different oral rehabilitations. *Open J. Stomatol.* **2012**, *2*, 21–22. [CrossRef]

24. Bakke, M. Bite force and occlusion. *Semin. Orthod.* **2006**, *12*, 120–126. [CrossRef]

25. Eng, C.M.; Lieberman, D.E.; Zink, K.D.; Peters, M.A. Bite force and occlusal stress production in hominin evolution. *Am. J. Phys. Anthr.* **2013**, *151*, 544–557. [CrossRef] [PubMed]

26. Aida, N.; Shinya, A.; Yokoyama, D.; Lassila, L.V.J.; Gomi, H.; Vallittu, P.K.; Shinya, A. Three-dimensional finite element analysis of posterior fiber-reinforced composite fixed partial denture Part 2: Influence of fiber reinforcement on mesial and distal connectors. *Dent. Mater. J.* **2011**, *30*, 29–37. [CrossRef]

27. Kuroda, S.; Shinya, A.; Vallittu, P.K.; Nakasone, Y.; Shinya, A. Effect of water temperature on cyclic fatigue properties of glass-fiber-reinforced hybrid composite resin and its fracture pattern after flexural testing. *J. Adhes. Dent.* **2013**, *15*, 19–26.

28. Papadogiannis, Y.; Lakes, R.S.; Palaghias, G.; Helvatjoglu-Antoniades, M.; Papadogiannis, D. Fatigue of packable dental composites. *Dent. Mater.* **2007**, *23*, 235–242. [CrossRef]

29. Lohbauer, U.; Frankenberger, R.; Krämer, N.; Petschelt, A. Strength and fatigue performance versus filler fraction of different types of direct dental restoratives. *J. Biomed. Mater. Res. B Appl. Biomater.* **2006**, *76*, 114–120. [CrossRef]

30. IBM Corp. *Released, IBM SPSS Statistics for Windows, Version 20.0*; IBM Corp.: Armonk, NY, USA, 2011.

31. Yanagida, H.; Tanoue, N.; Minesaki, Y.; Kamasaki, Y.; Fujiwara, T.; Minami, H. Effects of polymerization method on flexural and shear bond strengths of a fiber-reinforced composite resin. *J. Oral. Sci.* **2017**, *59*, 13–21. [CrossRef] [PubMed]

32. Garoushi, S.; Shinya, A. Fiber-reinforced onlay composite resin restoration: A case report. *J. Contemp. Dent. Pr.* **2009**, *10*, 104–110. [CrossRef]

33. Garoushi, S.; Yokoyama, D.; Shinya, A.; Vallittu, P.K. Fiber-reinforced composite resin prosthesis to restore missing posterior teeth: A case report. *Libyan. J. Med.* **2007**, *2*, 139–141. [CrossRef] [PubMed]

34. Lohbauer, U.; Frankenberger, R.; Krämer, N.; Petschelt, A. Time-dependent strength and fatigue resistance of dental direct restorative materials. *J. Mater. Sci. Mater. Med.* **2003**, *14*, 1047–1053. [CrossRef] [PubMed]

35. Selvaraj, A.; Nachimuthu, L.; Manickam, M. Effect of polyaramid reinforced sisal epoxy composites: Tensile, impact, flexural and morphological properties. *J. Mater. Res. Technol.* **2020**, *9*, 7947–7954.

36. Röhrle, O.; Saini, H.; Ackland, D.C. Occlusal loading during biting from an experimental and simulation point of view. *Dent. Mater.* **2018**, *34*, 58–68. [CrossRef]

37. Umesh, S.; Padma, S.; Asokan, S.; Srinivas, T. Fiber Bragg Grating based bite force measurement. *J. Biomech.* **2016**, *49*, 2877–2881. [CrossRef] [PubMed]

Article

Fracture Properties of Concrete in Dry Environments with Different Curing Temperatures

Zhengxiang Mi [1,2], Qingbin Li [2], Yu Hu [2,*], Chunfeng Liu [3] and Yu Qiao [3]

[1] School of Highway, Chang'an University, Xi'an 710064, China; mizxiang@chd.edu.cn
[2] State Key Laboratory of Hydroscience and Engineering, Tsinghua University, Beijing 100084, China; qingbinli@tsinghua.edu.cn
[3] China Three Gorges Projects Development Co., Ltd., Chengdu 610017, China; liu_chunfeng@ctg.com.cn (C.L.); qiao_yu@ctg.com.cn (Y.Q.)
* Correspondence: yu-hu@mail.tsinghua.edu.cn; Tel.: +86-010-62781161

Received: 12 June 2020; Accepted: 7 July 2020; Published: 9 July 2020

Abstract: This paper investigated the fracture properties of concrete in dry environments with different curing temperatures (5, 20, 40, and 60 °C). For each curing condition, the key fracture parameters of concrete were tested using wedge splitting specimens at five different ages (3, 7, 14, 28, and 60 d). The results show that in dry environments, the effective fracture toughness and fracture energy of concrete exposed to elevated temperatures increased at a relatively high growth rate at an early age. Nevertheless, the growth speed of effective fracture toughness and fracture energy decreased more quickly at elevated temperatures in the later stages. As a result, the concrete cured at higher temperature exhibited lower ultimate values of fracture parameters, and vice-versa. Namely, a temperature crossover effect was found in the effective fracture toughness and fracture energy of concrete under dry environments. Considering the early growth rate and ultimate values of fracture parameters, the optimum temperature suitable for concrete fracture properties development under dry condition was around 40 °C.

Keywords: concrete; fracture properties; dry environments; different curing temperatures; temperature crossover effect

---

1. Introduction

Concrete is used in a wide variety of structures, which are usually exposed to changing temperature and moisture content. The development of concrete mechanical and physical characteristics is significantly affected by the curing conditions. Harsh environmental conditions not only make the on-site casting and quality control process difficult but also accelerate setting, promote uneven distribution, reduce the later-age strength, increase the likelihood of cracking both before and after hardening and deteriorate the durability [1–5]. In order to construct the concrete structures in extreme weather conditions (such as extremely hot and dry conditions, cool and damp conditions), we must pay high attention to the concrete deterioration caused by harsh climate.

As we all know, the temperature is a very important factor that affects the concrete properties, and extensive studies have been conducted thus far; but the majority of those studies were related to the influence of temperature on concrete mechanical parameters, such as compressive strength [6], flexural-tensile strength [7], splitting-tensile strength [8], and Young's elastic modulus [9]. In general, temperature has a double effect on the development of concrete mechanical parameters. A higher curing temperature can significantly promote the development of concrete elastic modulus and strength in the early stages, but it also decreases the ultimate values of elastic modulus and strength. This behavior is called the temperature inversion phenomenon or temperature crossover effect [10]. Furthermore, apart from the Young's elastic modulus and strength, the elevated curing temperature will also deteriorate the microstructure and durability of concrete. For example, Sharp et al. [11] found that the microstructures of concrete exposed to 60 °C exhibited larger apparent porosity than those exposed to 10 °C after one year, which was a significant reason that the concrete cured at 60 °C exhibited the lower ultimate strengths. Jiang et al. [12] demonstrated that with the increasing of curing temperature, the autogenous shrinkage of concrete increases, and the cracking resistance of concrete becomes worse. Especially, under high-temperatures curing conditions, the resistance to chloride irons permeability was weakened [13]; further, after prolonged exposure, the concrete exposed to higher temperature showed anomalous expansion and accompanied by micro-cracks, leading to a serious deterioration of the concrete structure durability [14].

The humidity is another important environmental factor affecting the concrete properties, and the concrete is usually placed at reduced humidity in engineering practice, resulting in that the free water inside the concrete evaporates into the surrounding at lower humidity. An insufficient moisture supply owing to this evaporation and diffusion not merely worsens the hydration reaction and microstructure of concrete but also drastically deteriorates the mechanical and fracture properties [15]. For example, Flatt et al. [16] theoretically proved that C_3S stop hydrating when the internal relative humidity is less than 80%. Sagrario et al. [17] noticed that as the curing humidity decreases, the length of the chemical reaction chain forming the C-S-H gel increases, leading to a lower degree of hydration. Nawa et al. [18] reported that the strength at 60% relative humidity was 41% smaller than that in water after 7 d. Clearly, mechanical and physical properties of concrete measured under saturation conditions are larger than that of concrete under the realistic circumstance, and the influence of ambient humidity should be taken as seriously as that of temperature.

The fracture characteristics of concrete can dramatically affect the structural response of concrete members in the harsh environment, and the influence of ambient temperature and humidity on the fundamental fracture parameters of concrete should be studied. This is useful for a comprehensive understanding of the fracture properties of concrete exposed to realistic conditions, which is the key premise for assessing crack stability and predicting concrete structure failure based on fracture mechanics. However, so far, less information is available regarding the influence of environmental conditions on fracture properties of concrete, and some of the conclusions are contradictory. For instance, Buyukozturk et al. [19] clarified that due to the action of the pore water pressure inside the specimen, the fracture energy of concrete increases with reducing moisture. Mi et al. [20] observed that the value of fracture energy of concrete under dry conditions was smaller than that under saturated curing conditions. Besides, Yu et al. [21] obtained fracture energy gain data for concrete exposed to 14, 23, and 35 °C, and they clarified that concrete fracture energy improved dramatically with increasing temperature. Unfortunately, they did not investigate the fracture energy of concrete after 20 d. Analogously, Li et al. [22] observed that the fracture energy of full-graded concrete cast in hot weather was larger than that of concrete cast in cold weather from 28 d to 180 d. From the above results, it can be seen that fracture energy of concrete increased with increasing curing temperature. Conversely, Huang [23] clarified that the fracture energy of concrete stored at elevated temperature in summer was much smaller than that at 20 °C after 56 d, and the former was only half of the latter. This indicated that the elevated temperature deteriorated the fracture properties of concrete in later ages, which was in line with the results observed by Mi et al. [24]. Further, Mechtcherine [25] demonstrated that concrete fracture energy was less dependent on the temperature from 2 °C to 50 °C. It is clear that there is no consensus concerning the effect of ambient conditions on concrete fracture properties. What's more, there are very litter or no studies explaining the influence of temperature and humidity coupling on the fracture properties of concrete.

Therefore, the current paper presents the investigation of the fracture characteristics of concrete in dry environments with different curing temperatures. The fracture properties of concrete including fracture toughness and fracture energy were measured under a constant relative humidity of 50% with four different temperatures. The information about the fracture properties of concrete under hot-dry (or cold-dry) environments was not available in previous publications. The experimental data is helpful for a deeper and comprehensive understanding of the fracture properties of concrete under extreme climate environment, which can greatly enhance the numerical simulation of cracking behavior of concrete members in harsh environments and promote the development of cracking design specifications for concrete structures.

2. Experimental Program

2.1. Materials and Mix Proportions

The cement employed was Portland cement with the grade of P.I 42.5 satisfying the specifications of ASTM C150. The bulk density of cement equals 1450 kg·m^{-3}. The fly ash accounted for 25% of the cementitious materials, which is mainly for saving cement and reducing the heat of hydration. The chemical composition of cementitious materials was measured using XRD, and the results were given in Table 1. The coarse aggregate was crushed basalt, physical properties and gradation of which met the specifications of ASTM C33/C33M. The fine aggregate used was natural river sand, and its fineness modulus was 2.52. Further, the sand accounted for 39% of the total aggregate by weight, and this proportion remained unchanged. After comprehensively considering the fluidity and strength, the effective water to binder ratio of 0.4 and a 468 kg·m^{-3} cementitious content was selected. The concrete mix proportions were 1:0.40:1.50:2.34 (cementitious:water:sand:gravel). The slump of the concrete is 85 mm. Moreover, an external vibrator was utilized for concrete compaction.

Table 1. Chemical composition of cementitious materials (weight/%).

Composition	CaO	SiO$_2$	Al$_2$O$_3$	MgO	Na$_2$O	SO$_3$	Fe$_2$O$_3$	K$_2$O	Others
Cement	47.87	25.12	11.29	5.52	0.66	2.95	2.39	0.60	3.60
Fly ash	3.22	52.43	29.01	1.08	0.75	0.43	9.37	1.36	2.35

It is noted that before mixing, all the mixture constituents for concrete were reserved in the chambers with humidity and temperature previously set at 50% and 20 °C for not less than one day to ensure all the specimens having the same initial condition.

2.2. Curing Condition

In this study, the temperatures used were 5 °C, 20 °C, 40 °C, and 60 °C, corresponding to cold, normal, hot, and extremely hot weather, respectively. As for the curing humidity, it was set at 50% RH (relative humidity) for each temperature. The temperature and humidity during curing were controlled automatically using specialized equipment, and the control accuracy of temperature and relative humidity was ±2 °C and ±2% RH, respectively. Five test ages were designed for each curing condition, and three replicates fracture samples were poured for each age. The experimental data were evaluated and analyzed using averages of three identical samples. However, it should be emphasized that for a given variable, if the value of a specimen deviated from the average by more than 10%, then an additional specimen was conducted to improve the credibility of the results.

2.3. Specimen Preparation

The notched wedge splitting specimens are employed for investigating concrete fracture properties. The advantages of using it are that its self-weight has an insignificant effect on test results and the ratio of ligament length to sample size is considerably large [26]. However, to avert the unexpected failure of the sample and improve the stability of the sample, the double supports were used instead of the traditional sample with one-line support [27]. The test principle is shown in Figure 1a, where the vertical load is converted into the horizontal splitting force through a wedge-loading fixture. In the present test, the wedge angle is 15°.

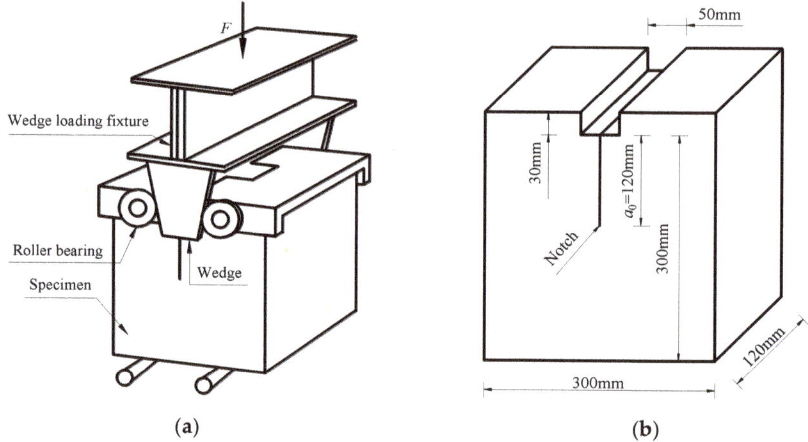

Figure 1. The fracture test specimen: (**a**) principle of wedge splitting test method; (**b**) the dimension of the specimen.

The dimension of the wedge splitting specimen was 120 × 300 × 300 mm (thickness × width × height), as shown in Figure 1b. The height of the specimen was greater than the recommended height in the specification in order to measure the fracture energy and fracture toughness independent of specimen size. The key purpose of this research is to investigate fracture properties of concrete under dry environments instead of the size effect, so a single specimen was adopted. In the concrete casting process, a steel plate was inserted into the tensile surface of the sample to form a notch. The thickness and initial length of the notch were 2 mm and 120 mm, respectively.

The specimen was poured in steel mold designed and manufactured according to experimental requirements, and each specimen was vibrated for about 1 min on a vibrator to eliminate air bubbles and to increase the density. After pouring, the sample was kept at natural conditions, and the pouring surface was covered with a layer of waterproof plastic film to prevent desiccation. Approximately 4.5 h after setting, the steel plate was cautiously pulled out. Subsequently, the sample was moved into the corresponding environmental box and kept there for 24 h. Afterward, the steel mold was removed and the concrete transferred into the same box again, where it was maintained until the test age. This process guarantees that the concrete has sufficient strength to avert damaging when placed, moved, and de-mold. Besides, this procedure also ensures that when concrete fracture properties begin to progress, all specimens achieve the same characteristics for better comparison. On the contrary, if the specimen is immediately transferred to the corresponding environment box after casting, the concrete already has different hydration degrees at the onset of its fracture properties development, resulting in difficulties in comparison [9]. Finally, this process is closer to practice, where concrete structures are usually cured for some time before being exposed to the working environment.

In addition to fracture specimens, additional cubes with a side length of 100 mm were cast using the same batch of concrete to determine the elastic modulus and strength for each research temperature. The cubes' size and test procedure are consistent with the specification of DL/T 5150-2001 [28]. Although the size of the specimens differs with that recommended by ASTM C 496, test specification conforms to the ASTM C 496 [29]. For each investigated curing condition, eight identical cubes were poured, four of which were used to measure the strength, and the remaining four were employed to evaluate the elastic modulus.

2.4. Testing

The wedge splitting specimens were tested according to ASTM E 1820 on an Instron servo-hydraulic testing machine. The fracture experiment set-up is shown in Figure 2. During loading, the main deformation of crack mouth opening displacement (CMOD) was monitored in real-time by a clip-gauge mounted at the mouth of the notch. The capacity and accuracy of the clip-gauge were 5 mm and ±0.2 μm, respectively. The tests were controlled by the CMOD. To start the test, the specimen was preloaded with 0.2 kN. Subsequently, the Instron testing machine was driven by an initial displacement criterion of 0.04 mm·min^{-1}. In the post-peak range, the loading rate was gradually increased up to 0.08 mm·min^{-1}. The Instron testing machine stops the loading when the load drops to 95% of the peak load. The main reason for choosing this loading scheme was to make the specimen failed within 30 to 40 min. During the test, the information on load and CMOD was collected twice per second. Due to the use of a testing machine with large stiffness, the crack propagated steadily during the test.

Figure 2. The fracture experiment set-up.

The elastic modulus and compressive strength at 28 d were measured using a universal testing machine with a range of 100 tons and a linear variable displacement transducer. To start the test, the sample was preloaded with 5 kN. Thereafter, the testing machine was driven by the displacement criterion of 0.2 mm·min^{-1}. The elastic modulus and compressive strength were determined by averaging the test results of the four identical cubes. Table 2 presents the measured values of elastic modulus and compressive strength under different curing conditions.

Table 2. Mechanical properties of concrete under different temperature at 28 d.

Curing Temperature	5 °C	20 °C	40 °C	60 °C
Compressive strength (MPa)	31.83	34.64	36.79	36.44
Elastic modulus (GPa)	23.36	26.72	29.26	28.88

3. Results

3.1. Load-CMOD Curves

Figure 3 presents the complete load-CMOD curves for concrete in a dry environment with different temperatures, which was used to obtain the peak load and calculate the fracture parameters such as fracture energy. Owing to the heterogeneity of concrete, the load-CMOD curves measured by the same set of specimens are different. For each investigated temperature and age, only the averaged curve of the same group of specimens was presented to avoid confusion. The overall load-CMOD curve is smooth, demonstrating that the unloading process was slow after the peak load and the test was carried out in a stable test state.

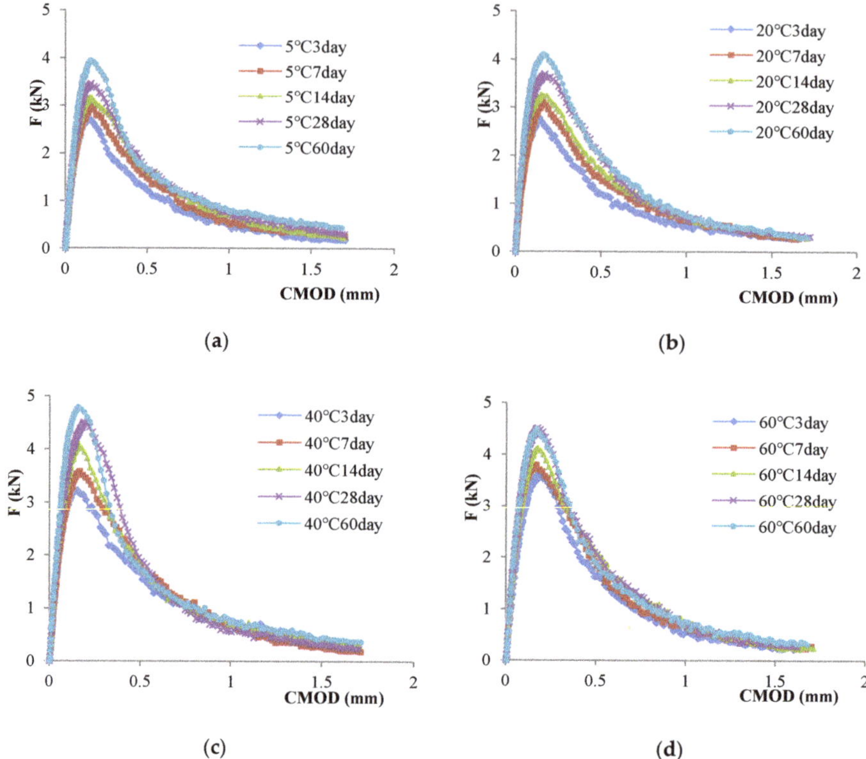

Figure 3. Load-crack mouth opening displacement (CMOD) curves under dry environments with different temperatures: (**a**) at 5 °C; (**b**) at 20 °C; (**c**) at 40 °C; (**d**) at 60 °C.

As presented in Figure 3, at the beginning of loading, there was a linear relationship between load and CMOD, indicating that concrete was in the stage of linear elasticity. The specimen did not undergo significant nonlinear deformation before the crack initiation or reaching the cracking load. For each curing condition, the initial slope increased as the concrete become older, which was attributed to the increase of the concrete strength and stiffness with time elapsing. Further, with the increase of curing time, the peak load grew markedly under each investigated curing temperature. For instance, for the specimens exposed to 5 °C, the peak load was 2.683 kN after 3 d, 2.906 kN after 7 d, 3.156 kN after 14 d, 3.427 kN after 28 d, and 3.911 kN after 60 d. The peak load increased by 14.1% from 28 d to 60 d, indicating that there was still a lot of un-hydrated cement inside the concrete under this curing condition. A similar phenomenon was also observed when the concrete kept at high temperature, but the rate of growth was lower than that of the concrete exposed to low temperature. Typically, for the concrete exposed to 60 °C, the peak load improved from 3.565 kN after 3 d of hydration to 4.491 kN after 60 d of hydration, where the peak load hardly increased beyond 28 d hydration.

Figure 3 also reveals that the peak load improved with the growing curing temperature during an early age, but this situation did not hold in later ages, with the greatest ultimate values of peak load corresponding to the specimens kept at 40 °C. In detail, at 3 d, the peak load improved from 2.683 kN at 5 °C to 2.723 kN at 20 °C, 3.366 kN at 40 °C, and finally to 3.565 kN at 60 °C with a final ascent of 32.9%. However, at 60 d, the peak load of concrete stored at 5 °C, 20 °C, 40 °C and 60 °C was 3.911 kN, 4.092 kN, 4.781 kN, and 4.491 kN, respectively. The peak load of concrete exposed to 60 °C was only 12.9% and 8.9% greater than that of concrete kept at 5 °C and 20 °C, respectively, and even was 6.5% smaller than that of concrete exposed to 40 °C.

Besides, the post-peak curvature of the curve, which represented the brittleness of the concrete, was related to the curing condition and age. This can be attributed to the improvement of the interfacial transition zone (ITZ), in which the stress concentration was decreased and more new hydration products were accumulated as the curing age was prolonged. Consequently, the bridging effect of the aggregate decreased and the fracture mode of concrete gradually transformed from the interfacial transition zone to aggregate being broken. Particularly, with further strengthening of ITZ at 60 d of curing age, the fracture path at high temperature was directly through the aggregate, which was similar to that of 28 d. As a result, the characteristic of load-CMOD curves in the unloading zone at 28 and 60 d followed the same pattern.

3.2. Fracture Energy

Fracture energy (G_F) represents the energy required to generate new cracks per unit area during crack propagation, which is applicable to energy release rate criterion. The simplest method to calculate fracture energy is the fracture work method. Thus, the fracture energy can be evaluated using Equation (1). Note that for a specimen that is not large, the effect of its weight is negligible. Thus, the own weight of the specimen was ignored when evaluating the fracture energy. Besides, the fracture experiment can never reach the zero load level [30]. Therefore, the far-tail constant value was usually used to determine the true fracture energy [31]. In the present paper, the fracture experiments did not stop before the unloading zone of the load-CMOD curve appeared a full tail. The part that was not measured was extended inversely by the fit-best expression in the evaluation of G_F.

$$G_F = \frac{W}{B(h - a_0)},\tag{1}$$

where W is the fracture work which equals the area below the measured splitting force against CMOD curve; a_0 is the length of pre-crack; B and h represent the thickness and height of wedge splitting specimen, respectively.

Figure 4 shows the fracture energy for concrete exposed to dry environments with different temperatures as a function of the age. In this figure, each point denotes the mean value of fracture energy evaluated by the three companion wedge splitting specimens, and the error bars are not presented to avoid displaying a pell-mell plot. In any case, the difference between one replicate and the mean of the same group was not more than 10% concerning the mean.

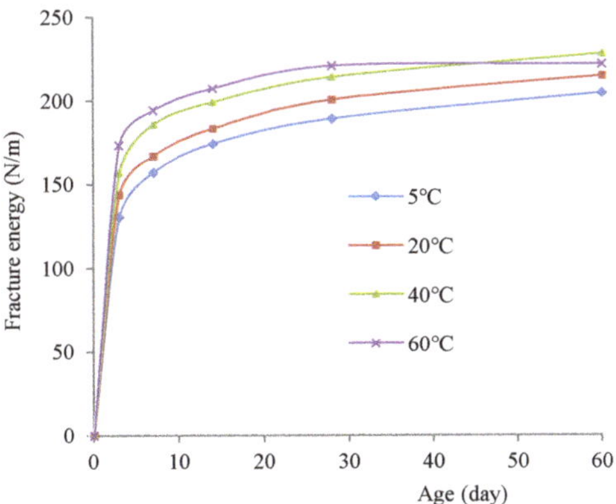

Figure 4. Fracture energy for concrete exposed to dry environments with different temperatures.

It can be seen from Figure 4 that the concrete fracture energy increased rapidly during the first 3 d, irrespective of the curing conditions investigated. Over time, fracture energy kept increasing at a reduced rate and gradually stabilized when the age was large enough. For instance, in the case of 5 °C, the fracture energy was 130.4 N·m^{-1} after 3 d of hydration, 157.2 N·m^{-1} after 7 d of hydration, 174.4 N·m^{-1} after 14 d of hydration, 189.4 N·m^{-1} after 28 d of hydration, and 204.7 N·m^{-1} after 60 d of hydration. The value of fracture energy at 60 d was 8.1% greater than that at 28 d. It means that the fracture energy continuously improved with the increase of age and the growth rate was still relatively great. In other words, in the case of 5 °C, 60 d of curing time were not sufficient to make the concrete achieve the stable state and there was a lot of un-hydrated cement inside the concrete. A similar phenomenon was observed under elevated temperature, but the fracture energy tended to reach a plateau value in a shorter time. Typically, for 60 °C, fracture energy was 173.7 N·m^{-1} after 3 d of hydration, 194.4 N·m^{-1} after 7 d of hydration, 207.7 N·m^{-1} after 14 d of hydration, 220.8 N·m^{-1} after 28 d of hydration, and 221.7 N·m^{-1} after 60 d of hydration. Obviously, in this case, the growth rate of fracture energy decreased significantly after 14 d and even stopped increasing after 28 d. At high curing temperature, the shorter stable time is mainly due to a higher temperature accelerating the hydration reaction rate of cement.

Figure 4 also reveals that in the early stage, the fracture energy of concrete increased significantly with the increasing temperature; but after 28 d of hydration, the curing temperature had the opposite effect on the fracture energy, that is, the larger later-age value of fracture energy was measured by concrete exposed to the lower temperature, and vice-versa. For instance, at 3 d, the fracture energy of concrete exposed to 60 °C was larger by 10.9%, 20.9%, and 32.9% compared to that of specimens exposed to 40 °C, 20 °C and 5 °C, respectively. However, for the specimens kept at elevated temperatures, the growth rate of the fracture energy decreased faster with the extension of curing time. As a result, the fracture energy difference between each temperature gradually decreased with time elapsing. In detail, after 28 d, the fracture energy at 60 °C reached 220.8 N·m^{-1}, which was 4.7%, 10.2%, and 16.6% higher than that of concrete at 40 °C, 20 °C and 5 °C, respectively. Further, after about 41 d of hydration, the fracture energy curves of concrete kept at 60 °C and 40 °C crossed each other, indicating that from 41-day onwards, the fracture energy of concrete exposed to 40 °C was larger than that of concrete exposed to 60 °C. Moreover, it can be easily inferred from the growth rate of fracture energy at 20 °C and 5 °C that, with the further increase of age, the curves at these two temperatures must intersect that at a higher temperature, and the values of fracture energy at these two temperatures certainly exceed that of concrete kept at the higher temperature. The maximum fracture energy limit was measured by concrete cured at 5 °C. Accordingly, the fracture energy of concrete cured at different temperatures has a temperature crossover effect. Considering the early growth rate and later-age value of fracture energy, around 40 °C was the optimum curing temperature propitious to the fracture energy development of concrete in dry environments. This optimal temperature was related to the faster hydration rate and higher hydration degree of the cementitious material, as well as to the microstructure of dense concrete with finer pores distribution [32].

3.3. Effective Fracture Toughness

Fracture toughness characterizes the cracking resistance of concrete, which is usually used for the stress intensity factor criterion. A smaller value of fracture toughness indicates that the concrete is apt to fracture abruptly before the occurrence of obvious unrecoverable deformation. The effective fracture toughness (K_{IC}) is calculated with the following equation [33]:

$$K_{IC} = \frac{F_{smax}}{B\sqrt{h}} \cdot \frac{3.675[1 - 0.12(\alpha - 0.45)]}{(1-\alpha)^{3/2}}, \quad \alpha = \frac{a_c}{h}, \tag{2}$$

where F_{smax} is the horizontally component of peak load; the meaning of h and B is the same as in Equation (1); a_c is the critical effective crack length, which is evaluated by the following equation [33]:

$$a_c = (h + h_0)\left(1 - \sqrt{\frac{13.18}{9.16 + CMOD_c \cdot E \cdot B / F_{smax}}}\right) - h_0, \tag{3}$$

where h_0 is the thickness of clip gauge holder; $CMOD_c$ is the crack opening displacement corresponding to F_{smax}; E is the tensile Young's modulus.

Figure 5 depicts the test results of effective fracture toughness against age under four curing conditions investigated. Although each point in Figure 5 denotes the mean values of the three replicate samples, the error bars are not shown again for the sake of brevity. Interestingly, the scattering of data observed in effective fracture toughness was smaller when compared with the case of fracture energy.

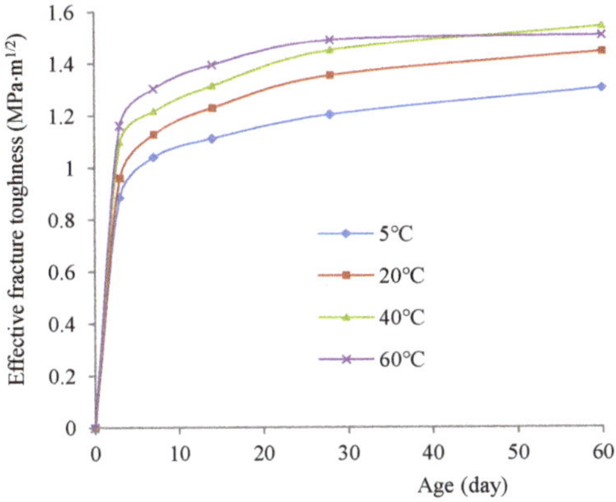

Figure 5. Effective fracture toughness for concrete exposed to dry environments with different temperatures.

During the unstable propagation, redundant cracks appeared and the propagation path of the main crack became more random; however, the energy consumed by these extra cracks was difficult to calculate, so these energies were ignored when evaluating fracture energy. As a result, the fracture energy produced larger randomness.

As can be presented in Figure 5, for each curing condition investigated, the effective fracture toughness was closely related to age and its value increased with age, which was the same as fracture energy with age. In the beginning, effective fracture toughness improved significantly with the growing age, because at this stage the reaction rate of cement hydration was quite fast. The growth rate increased by degrees and then reduced after maximizing. Subsequently, effective fracture toughness increased slightly with age at a reducing rate and finally reached a constant as the age was further extended. Typically, for 5 °C, the effective fracture toughness continuously increased from 0.889 MPam$^{0.5}$ after 3 d of hydration to 1.043 MPam$^{0.5}$ after 7 d of hydration, 1.113 MPam$^{0.5}$ after 14 d of hydration, 1.204 MPam$^{0.5}$ after 28 d of hydration, and finally 1.306 MPam$^{0.5}$ after 60 d of hydration, with a significant increase of 0.417 MPam$^{0.5}$ or 46.9%. Similarly, for 60 °C, the effective fracture toughness was 1.164 MPam$^{0.5}$ at the age of 3 d, 1.304 MPam$^{0.5}$ at the age of 7 d, 1.395 MPam$^{0.5}$ at the age of 14 d, 1.490 MPam$^{0.5}$ at the age of 28 d, and 1.509 MPam$^{0.5}$ at the age of 60 d. The fracture toughness at the age of 60 d was only 1.3% greater than that at the age of 28 d. This also demonstrated that it takes a shorter time for the effective fracture toughness to reach the constant value under higher temperatures.

Figure 5 also shows that, as observed in the fracture energy, there was also a temperature crossover effect in the effective fracture toughness of concrete in dry environments. As the curing temperature increased, concrete effective fracture toughness improved dramatically during the early ages, but this situation did not hold after 28 d, in which greater ultimate value of effective fracture toughness was acquired from the concrete kept at a lower temperature, and vice-versa. Specifically, at the age of 3 d, the fracture toughness of concrete kept at 60 °C was 1.164 MPam$^{0.5}$, which was 5.7% greater than at 40 °C, 21.1% greater than at 20 °C, and 30.9% greater than at 5 °C. Evidently, in the early days, higher temperatures promoted the rapid development of concrete effective fracture toughness. However, at the elevated temperature, the increasing rate of fracture toughness of concrete decreased faster with the extension of curing time. As a result, a larger later-age value of the effective fracture toughness was measured by concrete kept at a lower temperature. For example, for 60 °C, effective fracture toughness increased very litter beyond 14 d and almost reached a plateau value. In contrast, the specimens kept at a lower temperature (5 °C and 20 °C) had a higher increasing rate of effective fracture toughness. Representatively, at the age of 60 d, the fracture toughness of the concrete kept at 60 °C was only greater 15.5% and 4.4% when comparing concrete kept at 5 °C and 20 °C, respectively, and was smaller 2.3% when comparing concrete kept at 40 °C. In other words, the difference in the fracture toughness between different temperatures was decreasing as the curing age was prolonged. The fracture energy curves of concrete kept at 40 °C and 60 °C crossed each other at about 43 d, demonstrating that the effective fracture toughness of concrete cured at 40 °C was larger than that of concrete at 60 °C beginning from the 43rd day. Considering the increasing rate of effective fracture toughness at 5 °C and 20 °C, we can conclude that with the further extension of curing time, the curves at 5 °C and 20 °C will doubtlessly cross the curves at the higher temperature. This clearly showed that the optimum temperature propitious to the fracture energy development of concrete in dry environments was about 40 °C, which was the same as that for fracture energy.

4. Discussion

As mentioned earlier, during the first 3 d, the G_F and K_{IC} of concrete improved quickly with increasing age owing to the fast hydration rate of cement in this stage. The initial growth rate of concrete fracture parameters increased progressively and then reduced after maximizing, accompanying a steeper curve. As the curing temperature increased, the duration for this stage became shorter. In the later stage, the rate of hydration went down, resulting in that the concrete fracture parameters developed overtime at a smaller growth rate. In general, there should be the third stage, in which the fracture parameters of concrete tended to their stable values, as the age was further extended. However, at low temperatures, the third stage was not seen or was not notable because this study only investigated the age of up to 60 d, meaning that there was a lot of un-hydrated cement inside the concrete under these curing temperatures. This variation of concrete fracture parameters regarding the age was consistent with other independent investigations [34]. For example, Beygi et al. [35] found that with the increase of test age from 3 to 90 d, G_F and K_{IC} increase from 0.961 MPam$^{0.5}$ to 1.448 MPam$^{0.5}$ and from 99.7 N·m^{-1} to 126.5 N·m^{-1}, respectively. Lee et al. [36] demonstrated that the fracture energy shows a rapid increase during earlier and then starts to converge to a certain extent of 173.1 N·m^{-1} at 28 d. The main reason for the increase of concrete G_F and K_{IC} with age was the strengthening of the ITZ between the paste and aggregate. ITZ is the most sensitive area inside the concrete, where the largest number of micro-cracks are formed [37]. In other words, the failure of concrete usually depends on the strength of its weakest area. At an early age, there was a lot of un-hydrated cement in the ITZ, and it resulted in high porosity in this zone [38]. The bridging effect of the aggregate was very strong causing the cracks to propagate along with the interface, and a large number of aggregates were extracted from the matrix, as presented in Figure 6a. As the reaction of cement hydration progressed, an increasing number of hydration products accumulated in the ITZ and the ITZ pores were stuffed with hydration products, indicating that the size and content of pores in ITZ and cement paste decreased. The ITZ and cement paste also became stronger. As a result, the bridging effect of the aggregate became weaker and a growing number of aggregates were broken. The fracture mode under load also changed from around aggregate or bond zones through the aggregate directly, as presented in Figure 6b. Further, the aggregate strength was greater than that of the ITZ and cement paste, demonstrating that more energy was needed in order to get over the enhanced ITZ and cement paste and the aggregates. Consequently, the greater values of fracture energy and effective fracture toughness were obtained at later ages. However, as the strength of the ITZ and cement paste further improved, the fracture pattern did not alter dramatically, where an only increasing number of aggregates were extracted. Therefore, effective fracture toughness and fracture energy did not increase significantly, but gradually tended to their stable values.

(a)   (b)

Figure 6. The fracture surface of concrete at different ages: (**a**) at 3 d; (**b**) at 60 d.

More importantly, as is evident from the data in Figures 5 and 6, the effective fracture toughness and fracture energy of concrete cured at elevated temperatures developed at a faster speed at the early stage. For instance, after 3 d of hydration, the G_F and K_{IC} of concrete cured at 60 °C was 20.9% and 21.1% larger respectively than that of concrete cued at 20 °C. This promotion effect of temperature on the early fracture parameters of concrete is related to the accelerated hydration reaction rate of cement, and the finer pore distribution and denser microstructure [39]. Over time, however, the rate of growth of the fracture parameters of concrete kept at high temperatures also decreased at a higher rate, which resulted in these samples obtained the smaller ultimate values of G_F and K_{IC} than those of samples kept at reduced temperatures. For example, after 60 d of hydration, the difference in the G_F and K_{IC} between the samples exposed to 60 °C and 20 °C was reduced to 4.2% and 3.3%, respectively. Even the values of G_F and K_{IC} at 60 °C were less than that at 40 °C beginning from the 43rd day and 41st day, respectively. Further, the growth rate of fracture parameters at 5 °C and 20 °C also indicated that with the further extension of curing time, the curves at 5 °C and 20 °C will doubtlessly cross the curves at the higher temperature. Namely, although a higher temperature accelerated the rate of hydration reaction of cement and promoted the development of G_F and K_{IC} in the early stage, it harmed the development of the ultimate value of fracture parameters of concrete. The G_F and K_{IC} of concrete showed a temperature crossover effect, and the optimum temperature propitious to their development was approximately 40 °C. Such a trend was in line with previous observations found by other scholars with a slightly different method. Oswaldo et al. [40] claimed that curing at high temperatures was conducive to rapid strength gain, but at later ages, the greatest strength was measured by concrete cured at 20 °C. Boubekeur [41] found that the concrete compressive strength cured at 50 °C was 7% smaller than that of concrete stored at normal temperature after 28 d. Lee et al. [42] demonstrated a 15% reduction in the concrete compressive strength cured at 60 °C after 14 d compared to that cured at 40 °C. Nasir et al. [43] clarified that after 28 d the tensile strength improved with the pouring temperature up to 32 °C, but decreased these properties as a further increase in temperature.

The adverse effect of high-temperature on the later-age fracture properties of concrete makes us wonder: What happened at the microscopic level at a high-temperature? We concluded that this adverse effect of high-temperature on the effective fracture toughness and fracture energy of concrete was related to some different mechanisms. Firstly, this phenomenon is caused by the non-homogeneous dispersion of hydration products inside the pores of the hardened cement paste [44]. The hydration reaction of cement was accelerated at elevated temperature, so there was neither sufficient time for the hydration products to disperse surrounding nor sufficient time for them to precipitate and distribute evenly. Consequently, more and more hydrated products were concentrated surrounding the hydrating or un-hydrated particles of cement, resulting in a more non-homogeneous distribution of hydrated products and causing the concrete to become a porous structure [45]. Secondly, the change in the C-S-H apparent density at high-temperature was another important reason for the decrease of later-age values of concrete fracture parameters [46]. Gallucci et al. [47] clarified that there was a remarkable enhancement in the C-S-H apparent density in the situation of elevated temperature, which came from the different assembly of the C-S-H. Owing to this increase, the filling space of C-S-H was reduced, leading to that the microstructure of the concrete was much coarser and porous, which had a detrimental effect on the concrete fracture parameters. Other mechanisms that resulted in the temperature crossover effect were cumulative of compact and low infiltration hydration products surrounding the cement particles and the delayed diffusion of the hydrate in cement secondary hydration reaction [40]. Namely, the elevated temperature might have resulted in the rapid evaporation of uncombined water in the concrete, which restrained the further spread of cement and limited the progress of hydration reaction, thereby restricting the forming of more C-S-H and the development of fracture parameters. Finally, the deviation of thermal expansion coefficients among the concrete components was also related to the negative effect [48]. In the situation of high-temperature, the concrete would expand in volume. The deformation of each constituent of concrete is not equal because of the difference in thermal expansion coefficients, which will cause a great stress build-up in the air void concentration area or the ITZ. When this stress exceeds the matrix tensile strength, micro-cracks begin to emerge in weak areas, especially those at the interface of the cement paste and the aggregate [49]. What is worse, due to the capillary action, these micro-cracks will become an ideal moisture path and increase the mobility of moisture inside the concrete, which can generate a pressure of expansion of the water vapor due to swelling of water vapor more quickly than getting away from the sample, leading to further micro-cracking. As a result, the fracture mode of concrete at elevated temperature has altered due to higher porosity and micro-cracking, which means that a decreasing number of aggregates were broken, as presented in Figure 7. In short, all the above factors, in particular the high porosity, together deteriorated the properties of concrete at later-age and induced a reduction in the ultimate values of the breaking fracture and the fracture toughness of the concrete at high temperature. On the other hand, concrete stored under low temperature (5 °C and 20 °C) continuously underwent the favorable effect, with a very slow hydration rate, which left the dissolved ions enough time to diffuse and resulted in a more uniform hydration products distribution and lower coarse porosity [50].

(a)   (b)

Figure 7. The fracture surface of concrete at different temperatures: (**a**) at 5 °C; (**b**) at 60 °C.

5. Conclusions

In the current research, the fracture properties of concrete in dry environments with different temperatures were investigated using the wedge splitting specimens. Four different temperatures (5 °C, 20 °C, 40 °C, and 60 °C) were considered, which represented cold, normal, hot, and extremely hot weather, respectively. The complete load-CMOD curves of specimens were obtained at 3, 7, 14, 28, and 60 d. The key fracture parameters of concrete were also evaluated. The following conclusions can be drawn from the experimental data:

1. In a dry environment, the gain in fracture energy of the concrete strongly depended on the curing temperature. An elevated temperature was beneficial to the rapid development of concrete

fracture energy at an early age. However, this situation did not hold at later ages, in which a higher value of the fracture energy at a later age was obtained by concrete stored at lower temperatures, and vice versa. Generally, at 60 d, the fracture energy of concrete cured at 60 °C was only larger 7.7% and 3.2% when comparing concrete stored at 5 °C and 20 °C, respectively, and even was smaller 2.8% when comparing concrete stored at 40 °C. Considering the early growth rate and later-age value of fracture energy, the optimum curing temperature for fracture energy of concrete under dry condition was around 40 °C.

2. In the case of the dry condition, the effective fracture toughness improved significantly as the increase of curing temperature in the early stage, but the influence of temperature on the concrete fracture toughness reversed after 28 d, wherein a relatively small ultimate value of effective fracture toughness was measured by the samples stored at the higher temperature. The temperature crossover effect was observed in concrete effective fracture toughness. From the temperature investigated, the most suitable curing temperature for the development of effective fracture toughness of concrete in dry environments was still 40 °C.

3. For a given curing condition, both the effective fracture toughness and fracture energy improved speedily with age at the beginning. After the growth rate reached their maximum value, these two fracture parameters increased slightly with age at a reducing rate, and eventually tended to their stable values as the age was further extended. Further, the effective fracture toughness and fracture energy tended to stabilize in a shorter time with the increasing temperature.

It should be noted that the aforementioned conclusions are only valid for the concrete studied in this paper, and the general conclusions for other types of concrete need further study.

Author Contributions: Conceptualization, Z.M.; formal analysis, Z.M.; funding acquisition, Q.L. and Y.H.; investigation, Z.M.; methodology, Z.M.; project administration, C.L. and Y.Q.; resources, C.L. and Y.Q.; supervision, Y.H.; writing—original draft, Z.M.; writing—review and editing, Q.L. and Y.H. All authors have read and agreed to the published version of the manuscript.

Funding: This research was funded by National Natural Science Foundation of China (51979145, 51839007), the Fundamental Research Funds for the Central Universities, CHD (300102219305), the Open Research Fund Program of State Key Laboratory of Hydroscience and Engineering (sklhse-2020-D-01), and the Research Projects of China Three Gorges Corporation (Contract numbers: BHT/0806 and WDD/0427).

Conflicts of Interest: The authors declare no conflicts of interest.

References

1. Jennings, H.M.; Thomas, J.J.; Gevrenov, J.S.; Constantinides, G.; Ulm, F.J. A multi-technique investigation of the nanoporosity of cement paste. *Cem. Concr. Res.* **2007**, *37*, 329–336. [CrossRef]
2. Zhu, H.; Hu, Y.; Li, Q.B.; Ma, M. Restrained cracking failure behavior of concrete due to temperature and shrinkage. *Constr. Build. Mater.* **2020**, *244*, 118318. [CrossRef]
3. Liu, Y.B.; Presuel-Moreno, F. Effect of elevated temperature curing on compressive strength and electrical resistivity of concrete with fly ash and ground-granulated blast-furnace slag. *ACI Mater. J.* **2014**, *111*, 531–541.
4. Zhu, H.; Li, Q.; Ma, R.; Yang, L.; Hu, Y.; Zhang, J. Water-repellent additive that increases concrete cracking resistance in dry curing environments. *Constr. Build. Mater.* **2020**, *249*, 118704. [CrossRef]
5. Mi, Z.X.; Hu, Y.; Li, Q.B.; Gao, X.F.; Yin, T. Maturity model for fracture properties of concrete considering coupling effect of curing temperature and humidity. *Constr. Build. Mater.* **2019**, *196*, 1–13. [CrossRef]
6. Castellano, C.C.; Bonavetti, V.L.; Donza, H.A.; Irassar, E.F. The effect of w/b and temperature on the hydration and strength of blastfurnace slag cements. *Constr. Build. Mater.* **2016**, *111*, 679–688. [CrossRef]
7. Cakir, O.; Akoz, F. Effect of curing conditions on the mortars with and without GGBFS. *Constr. Build. Mater.* **2008**, *22*, 308–314. [CrossRef]
8. Shoukry, S.N.; William, G.W.; Downie, B.; Riad, M.Y. Effect of moisture and temperature on the mechanical properties of concrete. *Constr. Build. Mater.* **2011**, *25*, 688–696. [CrossRef]
9. Zhang, J.Y.; Cusson, D.; Monteiro, P.; Harvey, J. New perspectives on maturity method and approach for high performance concrete applications. *Cem. Concr Res.* **2008**, *38*, 1438–1446. [CrossRef]

10. Tanyildizi, H. Fuzzy logic model for prediction of mechanical properties of lightweight concrete exposed to high temperature. *Mater. Des.* **2009**, *30*, 2205–2210. [CrossRef]

11. Escalante-Garcia, J.I.; Sharp, J.H. The microstructure and mechanical properties of blended cements hydrated at various temperatures. *Cem. Concr. Res.* **2001**, *31*, 695–702. [CrossRef]

12. Jiang, C.J.; Yang, Y.; Wang, Y.; Zhou, Y.N.; Ma, C.C. Autogenous shrinkage of high performance concrete containing mineral admixtures under different curing temperatures. *Constr. Build. Mater.* **2014**, *61*, 260–269. [CrossRef]

13. Care, S. Effect of temperature on porosity and on chloride diffusion in cement pastes. *Constr. Build. Mater.* **2008**, *22*, 1560–1573. [CrossRef]

14. Patel, H.H.; Bland, C.H.; Poole, A.B. The microstructure of concrete cured at elevated temperature. *Cem. Concr. Res.* **1995**, *25*, 485–490. [CrossRef]

15. Michael, G.; Lars, W. A method based on isothermal calorimetry to quantify the influence of moisture on the hydration rate of young cement pastes. *Cem. Concr. Res.* **2010**, *40*, 867–874.

16. Flatt, R.J.; Scherer, G.W.; Bullard, J.W. Why alite stops hydrating below 80% relative humidity. *Cem. Concr. Res.* **2011**, *41*, 987–992. [CrossRef]

17. Aparicio, S.; Sagrario, M.R.; Miguel, M.A.; Jose, V.F. The effect of curing relative humidity on the microstructure of self-compacting concrete. *Constr. Build. Mater.* **2016**, *104*, 154–159. [CrossRef]

18. Saengsoy, W.; Nawa, T.; Termkhajornkit, P. Influence of relative humidity on compressive strength of fly ash cement paste. *J. Struct. Constr. Eng.* **2008**, *73*, 1433–1441. [CrossRef]

19. Lau, D.; Buyukozturk, O. Fracture characterization of concrete/epoxy interface affected by moisture. *Mech. Mater.* **2010**, *42*, 1031–1042. [CrossRef]

20. Mi, Z.X.; Hu, Y.; Li, Q.B.; An, Z.H. Effect of curing humidity on the fracture properties concrete. *Constr. Build. Mater.* **2018**, *169*, 403–413. [CrossRef]

21. Yu, B.J.; Ansari, F. Method and theory for nondestructive determination of fracture energy in concrete structures. *Aci Struct. J.* **1996**, *93*, 602–613.

22. Li, Q.B.; Guan, J.F.; Wu, Z.M.; Dong, W.; Zhou, S.W. Equivalent maturity for ambient temperature effect on fracture parameters of site-casting dam concrete. *Constr. Build. Mater.* **2016**, *120*, 293–308. [CrossRef]

23. Huang, Y.B.; Qian, J.S. The influence on high strength concrete mechanical properties by age and curing condition. *Bull. Chin. Ceram. Soc.* **2007**, *26*, 427–430.

24. Mi, Z.X.; Hu, Y.; Li, Q.B.; Zhu, H. Elevated temperature inversion phenomenon in fracture properties of concrete and its application in maturity model. *Eng. Fract. Mech.* **2018**, *199*, 294–307. [CrossRef]

25. Mechtcherine, V. Fracture mechanical behavior of concrete and the condition of its fracture surface. *Cem. Concr. Res.* **2009**, *39*, 620–628. [CrossRef]

26. Bretschneider, N.; Slowik, V.; Villmann, B.; Mechtcherine, V. Boundary effect on the softening curve of concrete. *Eng. Fract. Mech.* **2011**, *78*, 2896–2906. [CrossRef]

27. Zhao, Z.F.; Kwon, S.H.; Shah, S.P. Effect of specimen size on fracture energy and softening curve of concrete: Part, I. Experiments and fracture energy. *Cem. Concr. Res.* **2008**, *38*, 1049–1060. [CrossRef]

28. *DL/T 5150-2001, Test Code for Hydraulic Concrete*; China Electric Power Press: Beijing, China, 2006.

29. *ASTM C 496/C496M-17, Standard Test Method for Splitting Tensile Strength of Cylindrical Concrete Specimens*; ASTM International: West Conshohocken, PA, USA, 2017; pp. 1–5.

30. Lee, J.; Lopez, M.M. An experimental study on fracture energy of plain concrete. *Int. J. Concr. Struct. Mater.* **2014**, *8*, 129–139. [CrossRef]

31. Elices, M.; Guinea, G.V.; Planas, J. Measurement of the fracture energy using three-point bend tests: Part 3–Influence of cutting the P-δ tail. *Mater. Struct.* **1992**, *25*, 327–334. [CrossRef]

32. Kim, J.; Han, S.H.; Song, Y.C. Effect of temperature and aging on the mechanical properties of concrete Part I: Experimental results. *Cem. Concr. Res.* **2002**, *32*, 1087–1094. [CrossRef]

33. Xu, S.L.; Reinhardt, H.W. Determination of Double-K criterion for crack propagation in quasi-brittle fracture, Part III: Compact tension specimens and wedge splitting specimens. *Int. J. Fract.* **1999**, *98*, 179–193. [CrossRef]

34. Bella, C.D.; Michel, A.; Stang, H. Early age fracture properties of microstructurally-designed mortars. *Cem. Concr. Compos.* **2017**, *75*, 62–73. [CrossRef]

35. Beygi, M.H.A.; Kazemi, M.T.; Nikbin, I.M.; Amiri, J.V. The effect of aging on the fracture chracteristics and ductility of self-compacting concrete. *Mater. Des.* **2014**, *55*, 937–948. [CrossRef]

36. Lee, Y.; Kim, J.K. Fracture characteristics of concrete at early ages. *Int. J. Concr. Struct. Mater.* **2006**, *18*, 191–198.

37. Nezerka, V.; Bily, P.; Hrbek, V.; Fladr, J. Impact of silica fume, fly ash, and metakaolin on the thickness and strength of the ITZ in concrete. *Cem. Concr. Compos.* **2019**, *103*, 256–262. [CrossRef]

38. Elsharief, A.; Cohen, M.D.; Olek, J. Influence of aggregate size, water cement ratio and age on the microstructure of the interfacial transition zone. *Cem. Concr. Res.* **2003**, *33*, 1837–1849. [CrossRef]

39. Lothenbach, B.; Winnefeld, F.; Alder, C.; Wieland, E.; Lunk, P. Effect of temperature on the pore solution, microstructure and hydration products of Portland cement pastes. *Cem. Concr. Res.* **2007**, *37*, 483–491. [CrossRef]

40. Oswaldo, B.D.; Lauren, Y.Z.; Jose, I.E.G. Influence of the long term curing temperature on the hydration of alkaline binders of blast furnace slag-metakaolin. *Constr. Build. Mater.* **2016**, *113*, 917–926.

41. Boubekeur, T.; Ezziane, K.; Kadri, E.H. Estimation of mortars compressive strength at different curing temperature by the maturity method. *Constr. Build. Mater.* **2014**, *71*, 299–307. [CrossRef]

42. Lee, C.; Lee, S.; Nguyen, N. Modeling of compressive strength development of high-early-strength- concrete at different curing temperatures. *Int. J. Concr. Struct. Mater.* **2016**, *10*, 205–219. [CrossRef]

43. Nasir, M.; Al-Amoudi, O.S.B.; Al-Gahtani, H.J.; Maslehuddin, M. Effect of casting temperature on the strength and density of plain and blended cement concretes prepared and cured under hot weather conditions. *Constr. Build. Mater.* **2016**, *112*, 529–537. [CrossRef]

44. Tan, K.F.; John, M.N. Performances of concrete under elevated curing temperature. *J. Wuhan Univ. Technol. -Mater. Sci. Ed.* **2004**, *19*, 65–67.

45. Paul, M.; Glasser, F.P. Impact of prolonged warm (85°C) moist cure on Portland cement paste. *Cem. Concr. Res.* **2000**, *30*, 1869–1877. [CrossRef]

46. Sajedi, F.; Razak, H.A. Effects of curing regimes and cement fineness on the compressive strength of ordinary Portland cement mortars. *Constr. Build. Mater.* **2011**, *25*, 2036–2045. [CrossRef]

47. Gallucci, E.; Zhang, X.; Scrivener, K.L. Effect of temperature on the microstructure of calcium silicate hydrate (C-S-H). *Cem. Concr. Res.* **2013**, *53*, 185–195. [CrossRef]

48. Baoju, L.; Youjun, X.; Shiqiong, Z.; Jian, L. Some factors affecting early compressive strength of steam-curing concrete with ultrafine fly ash. *Cem. Concr. Res.* **2001**, *31*, 1455–1458. [CrossRef]

49. Yao, W.; Hu, X.; Liu, H.W.; Xia, K.W. Quantification of thermally induced damage and its effect on dynamic fracture toughness of two mortars. *Eng. Fract. Mech.* **2017**, *169*, 74–88. [CrossRef]

50. Komonen, J.; Penttala, V. Effect of high temperature on the pore structure and strength of plain and polypropylene fiber reinforced cement pastes. *Fire Technol.* **2003**, *39*, 23–34. [CrossRef]

Article

Chemical and Mechanical Roughening Treatments of a Supra-Nano Composite Resin Surface: SEM and Topographic Analysis

Francesco Puleio [1,*], Giuseppina Rizzo [1], Fabiana Nicita [1], Fabrizio Lo Giudice [1], Cristina Tamà [1], Gaetano Marenzi [2], Antonio Centofanti [1], Marcello Raffaele [3], Dario Santonocito [3] and Giacomo Risitano [3]

[1] Department of Biomedical and Dental Sciences and Morphofunctional Imaging, Messina University, 98100 Messina, Italy; rizzog@unime.it (G.R.); fabin92@hotmail.it (F.N.); fabrizio.logiudice@hotmail.it (F.L.G.); cristinatama6@gmail.com (C.T.); centofantiantonio@gmail.com (A.C.)

[2] Department of Neurosciences, Reproduction and Odontostomatological Sciences, University of Naples Federico II Via S. Pansini 5 80131 Napoli (NA), Italy; gaetano.marenzi@unina.it

[3] Engineering Department, Messina University, 98100 Messina, Italy; mraffaele@unime.it (M.R.); dsantonocito@unime.it (D.S.); grisitano@unime.it (G.R.)

* Correspondence: francesco.puleio@live.it; Tel.: +39-346-494-9761

Received: 21 May 2020; Accepted: 26 June 2020; Published: 28 June 2020

Abstract: Background: Repairing a restoration is a more advantageous and less invasive alternative to its total makeover. The aim of this study was to analyze the effects of chemical and mechanical surface treatments aimed at increasing the roughness of a supra-nano composite resin. Methods: 27 cylindrical blocks of microhybrid composite were made. The samples were randomly divided into nine groups (n = 3). The samples' surface was treated differently per each group: acid etching (35% H_3PO_4, 30 s and 60 s), diamond bur milling, sandblasting and the combination of mechanical treatment and acid etching. The samples' surface was observed by a scanning electron microscope (SEM) and a confocal microscope for observational study, and surface roughness (R_a) was recorded for quantitative analysis. Results: The images of the samples sandblasted with Al_2O_3 showed the greatest irregularity and the highest number of microcavities. The surfaces roughened by diamond bur showed evident parallel streaks and sporadic superficial microcavities. No significant roughness differences were recorded between other groups. The difference in roughness between the control group, diamond bur milled group and sandblasted group was statistically significant. ($p < 0.01$). Comparison between the diamond bur milled group and the sandblasted group was also significant ($p < 0.01$). Conclusion: According to our results, sandblasting is the best treatment to increase the surface roughness of a supra-nano composite.

Keywords: surface roughness; microhybrid composite; sandblasting; surface treatment; composite repair; minimal invasive dentistry

---

1. Introduction

Over the past few years, the quality of direct and indirect restorations has improved in terms of the adhesion strength, longevity and composition of the resin matrix and filler [1].

However, like all dental materials, the composites undergo deterioration processes as a consequence of mechanical (cyclic fatigue and wear) or thermal stresses and chemical degradation (enzymatic, hydrolytic and acidic) [2,3].

Fractures, marginal bacterial infiltration, dentin treatments, teeth or restoration color changes, indirect restorations or endodontic post detachment can also compromise the result and require a makeover [4–9].

In these cases, complete replacement can be a long and expensive procedure with the possibility of further healthy dental tissue loss and an increased risk of pulp exposure. Therefore, repairing the restoration by preserving parts or rebonding an indirect restoration can be an advantageous alternative as these techniques are less invasive and allow the prolongation of the efficiency of conservative therapies over time [5,10–16].

The prognosis of a repair and maintenance treatment depends on the adhesion strength achieved between the old restoration and the new composite material layer [10].

In clinical practice, the adhesion techniques and mechanisms at the dental tissues–composite or composite–composite interfaces are different [17].

The adhesion strength between two composite surfaces depends on their chemical composition and characteristics such as the roughness, conditioning procedures and ability to become wet of the polymerized surface [10].

Furthermore, contrary to what happens in the composite layering technique, during reparation, the material integration process is hindered by the relative lack of unpolymerized monomers [18–20].

After intraoral photopolymerization, the conversion rate from monomer to polymer is between 45% and 70% because not all monomers participate in the polymerization and the remaining monomers are available for new adhesion [21–23].

Considering that the number of unsaturated double bonds in monomers decrease over time, the adhesion strength reduces by between 25% and 80%, and consequently the effectiveness of the repair process is time-related [18–20].

The adhesion of new composite to an old restoration is achieved with a new bond established by residual monomers and micromechanical retention that exploits the surface irregularities of the old restoration.

In order to improve this type of adhesion, various surface treatments have been described in the literature such as surface roughening with diamond burs of different granulometry, sandblasting with aluminum oxide or silica sand, acid etching and the application of hydrogen peroxide or silane.

A systematic review of the data still appears insufficient to indicate the best method for repairing Bis-GMA-based resins [11,14,18,21,22]. The aim of this study is to analyze the effects of chemical and mechanical surface treatments and their combination on the roughness of a supra-nano composite resin surface.

2. Materials and Methods

27 cylindrical blocks (height 4 mm and diameter 6 mm) of supra-nano composite col. A1 (Estelite Sigma Quick, Tokuyama Dental, Japan) were made using a silicone mold matrix.

Considering a power level of 85% with a type-I error = 0.05, for parameter roughness (Ra), three samples for each independent group (surface-treated and independent control) were necessary.

The technical characteristics of the composite resin provided by the manufacturer were: a resin matrix (bisphenol A-glycidyl methacrylate (Bis-GMA) and triethylene glycol dimethacrylate (TEGDMA)) and particles of reinforcement (71% of the total volume) formed by spherical particles of silica and zirconia, with sizes ranging from 0.1 to 0.3 µm (average size of 0.2 µm).

The cylindrical blocks were obtained through two vertical composite increments of 2 mm inside the silicone matrix. Using a LED curing lamp (Valo, Ultradent, South Jordan, UT, USA) the composite layers were light-cured for 20″ at a distance of 1 mm at 3.200 mW/cm². In order to prevent the inhibition of polymerization due to the presence of oxygen, and to create a homogeneous surface, the last layer of composite was covered by a glass plate before light-curing.

To make the samples' surface uniform, all the composite blocks were polished under 4× magnification (EyeMag Pro S, Zeiss, Oberkochen, Germania), using Soft-Lex (3M ESPE, St Paul, Minnesota) coarse-grained, medium, fine and superfine discs for 10 s each. After each step the samples were washed with distilled water and air dried.

The samples were randomly divided into 9 groups (n = 3). The surface of the samples from each group was treated with different roughening protocols:

A) Control group, no surface treatment.
B) Etching for 30 s.
C) Etching for 60 s.
D) Roughening with diamond bur.
E) Roughening with diamond bur and etching for 30 s.
F) Roughening with diamond bur and etching for 60 s.
G) Sandblasting.
H) Sandblasting and etching for 30 s.
I) Sandblasting and etching for 60 s.

Etching was performed with 35% orthophosphoric acid (Ultra-Etch, Ultradent) for 30 s in groups B, E and H or 60 s in groups C, F and I. After etching, each sample was washed with distilled water and air dried.

Milling was carried out with a diamond bur with granulometry 151 μm (6837 KR Komet) mounted on a handpiece for 3 s (groups D, E and F)

Sandblasting (20 s) was performed with an intraoral sandblaster using Al_2O_3 50 μm (Dentoprep, Rønvig Dental).

The observational analysis of the treated surface of the samples was carried out with an SEM (Phenom G2, Phenom, Eindhoven, the Netherlands), at 2.100x magnification, with 5 kV voltage and secondary electrons (SE). This SEM does not require any treatment of the sample surfaces [24,25].

The samples were then observed and analyzed with a confocal microscope (Leica DCM 3D, Leica Microsystems) in order to measure roughness. In each sample an area of 2.5 mm^2 was selected using a systematic random sampling protocol for stereological and morphometrical analysis and the roughness (Ra) was calculated using the following formula:

$$Ra = \frac{1}{l} \int_0^l |Z(x)| dx$$

For each experimental group the mean and quantitative parameter standard deviation (SD) were calculated.

Data were analyzed using the Student's t-Test and differences of $p < 0.05$ were considered statistically significant.

The statistical analysis was performed using the SPSS 17.0 for Windows package and the Prism software package (GraphPad, La Jolla, CA, USA).

3. Results

3.1. SEM Analysis

The SEM observation showed that the different types of roughening protocol exhibited no substantial differences among samples of the same group.

Comparisons between groups showed different surface morphologies, displayed in Figure 1.

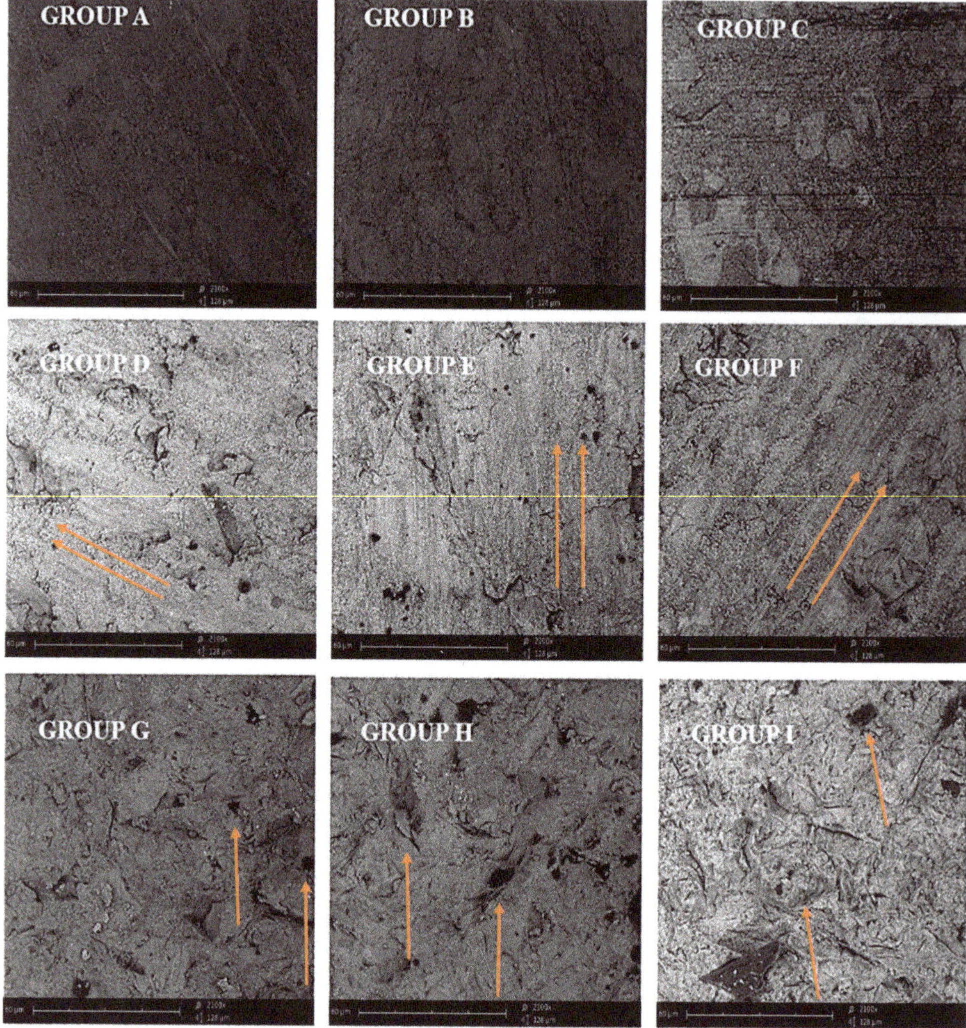

Figure 1. SEM images (2100×). Voltage: 5kV. Type of electron: secondary electron (SE). Groups A, B and C show a surface without irregularities; groups D, E and F show evident parallel streaks resulting from the bur action (arrows indicates the parallel streaks); groups G, H and I, the sandblasted samples, show the highest number of microcavities (arrows indicate microcavities).

The surface of the control group samples (group A) was homogeneous, without irregularities and without microcavities. Etching with 35% orthophosphoric acid for 30 s or 60 s (groups B and C) did not cause any observational modification of the composite surface. The surface roughened by diamond bur (group D) showed evident parallel streaks resulting from the bur action. The images of the sandblasted samples (group G) show greater irregularity and a high number of microcavities. Etching using 35% orthophosphoric acid for 30 s and 60 s (groups E, F, H and I) as an additional treatment to milling and sandblasting caused no change on the sample surface.

In all experimental groups, the optical confocal microscope analysis shows three-dimensional topographic images similar to the SEM observation (Figure 2).

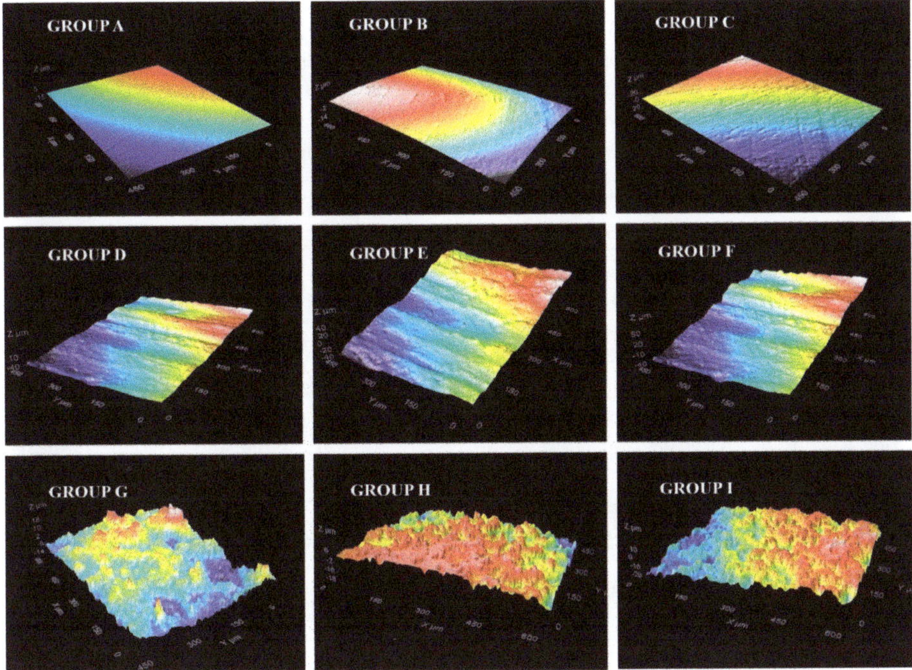

Figure 2. Confocal microscope. Three-dimensional surface topography. Red expresses the peaks; blue shows the depressions. Groups A, B and C show a surface with low irregularities; groups D, E and F show a moderate presence of surface irregularities; groups G, H and I show the highest presence of surface irregularities.

3.2. Profilometric Analysis

The mean values of roughness (R_a) and the standard deviation (SD) for each experimental group are shown in Figure 3.

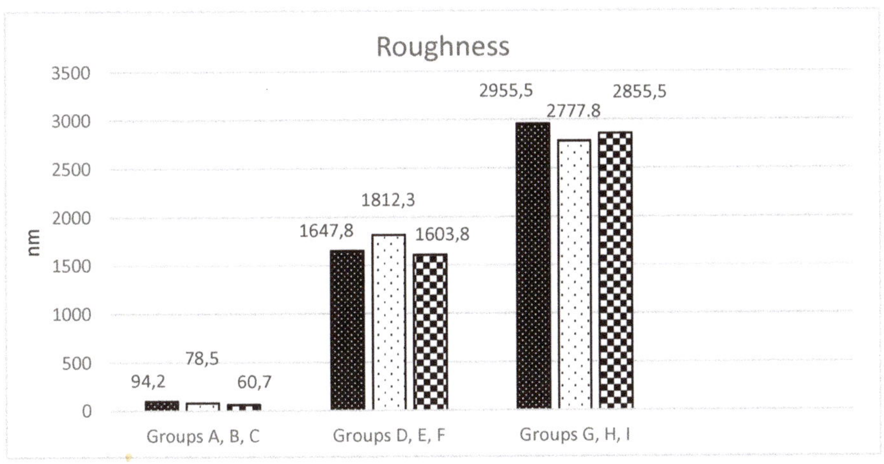

Figure 3. Roughness (Ra) mean values of experimental group.

There were no statistically significant differences in R_a between groups A (94.2 ± 44.7 nm), B (78.6 ± 23.3 nm) and C (60.7 ± 35.3 nm).

The same result was obtained by comparing the values for groups D (1647.8 ± 471.9 nm), E (1812.3 ± 300.7 nm) and F (1603.8 ± 280.4 nm), and those for groups G (2955.6 ± 572.9 nm), H (2777.8 ± 447.6 nm) and I (2855.6 ± 494 nm).

The comparison of R_a between group A (control) and groups D and G indicated statistically significant differences ($p < 0.01$). Additionally, the difference in R_a between groups D and G was statistically significant ($p < 0.01$).

4. Discussion

The introduction into clinical practice of build-up and indirect restorations made of composite resins requires the knowledge of adhesion mechanisms, especially when the surfaces are made of already polymerized composite [26]. This analysis can also be linked to the necessity of following a conservative approach in order to increase the longevity of direct restorations and the possibility of further repairing processes [27]. The treatments used most often to enhance the adhesive's action are chemical and mechanical surface roughening [14,18,19,21,26]. In this study, two mechanical and one chemical technique of surface conditioning and their combination were used. The mechanical techniques consisted of surface roughening by the action of a diamond bur or sandblasting with Al_2O_3. The chemical conditioning technique instead consisted of an etching procedure on the samples' surface with 35% orthophosphoric acid for 30 s or 60 s [28].

The images obtained with SEM and three-dimensional topography with an optical confocal microscope in all experimental groups agree, showing that among samples of the same groups the surface treatments cause substantially comparable morphological alteration. This highlights how the response of the same type of material to various techniques does not change and is therefore predictable.

The control group (treated exclusively by polishing with abrasive discs) presented a smooth, homogeneous surface without any irregularities or microcavities.

The surfaces roughened by a bur were irregular, with a limited number of microcavities distant from each other and with parallel streaks resulting from the action of the bur. The sandblasted samples showed greater irregularity and a high number of microcavities.

The comparison between the observations for these latter groups, according to previous studies, confirms the hypothesis that sandblasting with Al_2O_3 is the most suitable treatment for increasing the micromechanical retention of composite surfaces [29].

In our research, the treatment of etching alone, regardless of its duration, does not cause a significant improvement of composite roughness. Furthermore, there are not any substantial modifications on the surface even when the etching action is carried out on a previously mechanically conditioned surface. This outcome is independent of the duration of acid application. The observational data, therefore, seem to exclude any positive impact of the etching, according to Loomans et al., who stated that 35% orthophosphoric acid is not able to cause significant alterations to the resin filler. This component can be modified only by more aggressive acids, such as hydrofluoric acid [28].

The morphological findings (SEM and confocal) are confirmed by the profilometric analysis quantifying the average roughness (R_a) of the experimental groups.

The profilometric analysis (Figure 2) shows variable roughness among the experimental groups. The color scale from red to blue expresses the difference in height at different points of the same sample. To understand these data it is important to note that it is almost impossible to obtain a perfectly flat sample by hand, and this is particularly evident in the A, B and C groups, where the analysis shows the lowest R_a values but red and blue areas are visible. This color expression afforded the researchers a visual way of understanding the surfaces' pattern and without comparing their height. Groups A, B and C obtained the lowest R_a values; moreover, there were no significant differences in the average roughness between surfaces treated with etching only and the control group. Even when the effect of the etching on milled or sandblasted surfaces was evaluated, the variation in roughness was not statistically significant. The concordance between the observations shows how chemical treatment does not change the roughness of a smooth or previously mechanically roughened surface.

The average roughness increased significantly when comparing the control group to the samples (D, E and F) whose surface was roughened by the action of a diamond bur. This statistical evaluation was also valid for the sandblasted groups (G, H and I).

The comparison between the groups treated with mechanical conditioning shows that the sandblasted samples had significantly higher average roughness. These data agree with the observational findings and confirm that sandblasting is the best treatment possible for roughening a composite surface [28].

SEM observation allows us to directly observe the surface morphology to understand the modification related to different tools and the combination of materials and techniques [30]. The principal advantage of this technique is the possibility to directly observe samples without any major surface modification, especially samples with irregular and reflective surfaces, where producing bi-dimensional images does not give any information regarding surface roughness. In particular, the SEM used in this study gives a precise and reliable surface image not requiring any surface metallization prior to observation. The confocal laser microscope with profilometric analysis gives a 3D objective evaluation of the surfaces, providing a visual color scale that shows differences in height among different points of the sample and so provides quantitative data on the sample microsurface [31]. The combination between SEM analysis and the confocal laser provides the possibility to obtain and order numerical categories such as the smoothness, roughness and waviness of irregular and reflective surfaces along with a precise and detailed bi-dimensional image [32].

The main limitation of this study is related to the experimental samples that evaluated just one supra-nano composite. There is a need to compare the efficacy of this surface treatment on other composites, considering that different composite materials with different physical-chemical characteristics might respond to surface treatments in a specific manner.

In our experimental sample the specimens were not subjected to thermocycling procedures due to the in vitro experimental setting of the research. The thermocycling process gives the researchers the possibility to perform tests and evaluate the samples in a setting more similar to the clinical world. In the literature it has been described how aged composite shows lower adhesion values when compared to non-aged composite, probably due to hydrolytic degradation in the resin matrix that occurs in the oral environment, and due to the reduction of free radicals available that can react chemically with a fresh composite [20,33,34].

Moreover, further study is necessary to evaluate the bond strength of the composite surface and its effect.

5. Conclusions

Considering the results of our research, it is possible to conclude that among the treatments used to increase the roughness of a supra-nano composite surface, the most effective treatments are mechanical ones. Within this category of conditioning, sandblasting creates the best environment for the microretention of the adhesive system.

The use of orthophosphoric acid does not result in efficient surface roughness: the SEM images show a different appearance after 60 s acid etching; however, there is no statistically significant difference. According to our results, it can be concluded that sandblasting should be carried out to increase the surface roughness of an already polymerized supra-nano composite.

This treatment makes it possible to integrate the chemical forces generated between the monomers contained both in the adhesive and in the old composite surface with mechanical microretention.

Author Contributions: Conceptualization, F.P.; methodology, G.R.; software, M.R., D.S.; formal analysis, C.T., F.N.; investigation, G.M., A.C.; data curation, F.L.G.; writing—review and editing, F.P.; supervision G.R. All authors have read and agreed to the published version of the manuscript.

Funding: This research received no external funding.

Conflicts of Interest: The authors declare no conflict of interest.

References

1. Wiegand, A.; Stawarczyk, B.; Buchalla, W.; Tauböck, T.T.; Özcan, M.; Attin, T. Repair of silorane composite—Using the same substrate or a methacrylate-based composite. *Dent. Mater.* **2012**, *28*, e19–e25. [CrossRef]

2. da Rosa Rodolpho, P.A.; Cenci, M.S.; Donassollo, T.A.; Loguercio, A.D.; Demarco, F.F. A clinical evaluation of posterior composite restorations: 17-Year findings. *J. Dent.* **2006**, *34*, 427–435. [CrossRef] [PubMed]

3. Loomans, B.A.; Cardoso, M.V.; Roeters, F.J.; Opdam, N.J.; De Munck, J.; Huysmans, M.C.; Van Meerbeek, B. Is there one optimal repair technique for all composites. *Dent. Mater.* **2011**, *27*, 701–709. [CrossRef] [PubMed]

4. JafarzadehKashi, T.S.; Erfan, M.; Rakhshan, V.; Aghabaigi, N.; TabaTabaei, F.S. An in vitro assessment of the effects of three surface treatments on repair bond strength of aged composites. *Open Dent.* **2011**, *36*, 608–617.

5. Opdam, N.J.; van de Sande, F.H.; Bronkhorst, E.; Cenci, M.S.; Bottenberg, P.; Pallesen, U.; Gaengler, P.; Lindberg, A.; Huysmans, M.C.; van Dijken, J.W. Longevity of posterior composite restorations: A systematic review and meta-analysis. *J. Dent. Res.* **2014**, *93*, 943–949. [CrossRef] [PubMed]

6. Lo Giudice, R.; Pantaleo, G.; Lizio, A.; Romeo, U.; Castiello, G.; Spagnuolo, G.; Lo Giudice, G. Clinical and spectrophotometric evaluation of LED and laser activated teeth bleaching. *Open Dent. J.* **2016**, *10*, 242–250. [CrossRef]

7. Mohammadi, Z.; Giardino, L.; Palazzi, F.; Shalavi, S.; Alikhani, M.Y.; Lo Giudice, G.; Davoodpour, N. Effect of sodium hypochlorite on the substantivity of chlorhexidine. *Int. J. Clin. Dent.* **2013**, *6*, 173–178.

8. Lo Giudice, R.; Puleio, F.; Verrusio, C.; Matarese, M.; Alibrandi, A.; Lizio, A. Bulk vs. wedge shape layering techniques in V class cavities: Marginal infiltration evaluation. *G. Ital. Endod.* **2017**, *31*, 73–77. [CrossRef]

9. Lo Giudice, R.; Lipari, F.; Puleio, F.; Alibrandi, A.; Lo Giudice, F.; Tamà, C.; Sazonova, E.; Lo Giudice, G. Spectrophotometric evaluation of enamel color variation using infiltration resin treatment of white spot lesions at one year follow-up. *Dent. J.* **2020**, *8*, 35. [CrossRef]

10. Özcan, M.; Pekkan, G. Effect of different adhesion strategies on bond strength of resin composite to composite-dentin complex. *Oper. Dent.* **2013**, *38*, 63–72. [CrossRef]

11. Hamano, N.; Chiang, Y.C.; Nyamaa, I.; Yamaguchi, H.; Ino, S.; Hickel, R.; Kunzelmann, K.H. Effect of different surface treatments on the repair strength of a nanofilled resin-based composite. *Dent. Mater. J.* **2011**, *30*, 537–545. [CrossRef] [PubMed]

12. Hasan, N.H. The influence of Er: YAG Laser; Aluminum ox-ide and diamond bur on surface treatment of aged composite resin to repair restoration. *Al–Rafidain Dent. J.* **2012**, *12*, 257–265. [CrossRef]

13. Rossato, D.M.; Bandeca, M.C.; Saade, E.G.; Lizarelli, R.F.Z.; Bagnato, V.S.; Saad, J.R.C. Influence of Er:YAG laser on surface treatment of aged composite resin to repair restoration. *Laser Phys.* **2009**, *19*, 2144–2149. [CrossRef]

14. Fawzy, A.S.; El-Askary, F.S.; Amer, M.A. Effect of surface treatments on the tensile bond strength of repaired water-aged anterior restorative micro-fine hybrid resin composite. *J. Dent.* **2008**, *36*, 969–976. [CrossRef] [PubMed]

15. Fernández, E.; Martín, J.; Vildósola, P.; Oliveira Junior, O.B.; Gordan, V.; Major, I.; Bersezio, C.; Estay, J.; De Andrade, M.F.; Moncada, G. Can repair increase the longevity of composite resins? Results of a 10-year clinical trial. *J. Dent.* **2015**, *43*, 279–286. [CrossRef] [PubMed]

16. Fernández, E.M.; Martin, J.A.; Angel, P.A.; Mjör, I.A.; Gordan, V.V.; Moncada, G.A. Survival rate of sealed, refurbished and repaired defective restorations: 4-year follow-up. *Braz. Dent. J.* **2011**, *22*, 134–139. [CrossRef]

17. Lo Giudice, G.; Lizio, A.; Lo Giudice, R.; Centofanti, A.; Rizzo, G.; Runci, M.; Alibrandi, A.; Cicciù, M. The effect of different cleaning protocols on post space: A SEM study. *Int. J. Dent.* **2016**, *2016*, 1907124. [CrossRef]

18. Lucena-Martín, C.; González-López, S.; Navajas-Rodríguez de Mondelo, J.M. The effect of various surface treatments and bonding agents on the repaired strength of heat-treated composites. *J. Prosthet. Dent.* **2001**, *86*, 481–488. [CrossRef]

19. Özcan, M.; Corazza, P.H.; Marocho, S.M.; Barbosa, S.H.; Bottino, M.A. Repair bond strength of microhybrid, nanohybrid and nanofilled resin composite: Effect of substrate resin type, sur-face conditioning and aging. *Clin. Oral Invest.* **2013**, *17*, 1751–1758. [CrossRef]

20. Ozcan, M.; Barbosa, S.H.; Melo, R.M.; Galhano, G.A.; Bottino, M.A. Effect of surface conditioning methods on the micro-tensile bond strength of resin composite to composite after aging conditions. *Dent. Mater.* **2007**, *23*, 1276–1282. [CrossRef]

21. Rinastiti, M.; Ozcan, M.; Siswomihardjo, W.; Busscher, H.J. Immediate repair bond strengths of microhybrid, nanohybrid and nanofilled composites after different surface treatments. *J. Dent.* **2010**, *3*, 29–38. [CrossRef] [PubMed]

22. Rodrigues, S.A.; Ferracane, J.L., Jr.; Della Bona, A. Influence of surface treatments on the bond strength of repaired resin composite restorative materials. *Dent. Mater.* **2009**, *25*, 442–451. [PubMed]

23. Trujillo, M.; Newman, S.M.; Stansbury, J.W. Use of near-IR to monitor the influence of external heating on dental composite photopolymerization. *Dent. Mater.* **2004**, *20*, 766–777. [CrossRef] [PubMed]

24. De Ponte, F.S.; Favaloro, A.; Nastro Siniscalchi, E.; Centofanti, A.; Runci, M.; Cutroneo, G.; Catalfamo, L. Sarcoglycans and integrins in bisphosphonate treatment: Immunohistochemical and scanning electron microscopy study. *Oncol. Rep.* **2013**, *30*, 2639–2646. [CrossRef] [PubMed]

25. Giudice, R.L.; Rizzo, G.; Centofanti, A.; Favaloro, A.; Rizzo, D.; Cervino, G.; Squeri, R.; Costa, B.G.; Fauci, V.L.; Lo Giudice, G. Steam sterilization of equine bone block: Morphological and collagen analysis. *BioMed. Res. Int.* **2018**, *2018*, 9853765. [CrossRef]

26. Eliasson, S.T.; Tibballs, J.; Dahl, J.E. Effect of different surface treatments and adhesives on repair bond strength of resin composites after one and 12 months of storage using an improved microtensile test method. *Open Dent.* **2014**, *39*, E206–E216. [CrossRef]

27. Lo Giudice, R.; Lizio, A.; Cervino, G.; Nicita, F.; Puleio, F.; Ausiello, P.; Cicciù, M. The horizontal root fractures. Diagnosis, clinical management and three-year follow-up. *Open Dent. J.* **2018**, *12*, 687–695. [CrossRef]

28. Loomans, B.A.; Cardoso, M.V.; Opdam, N.J.; Roeters, F.J.; De Munck, J.; Huysmans, M.C.; Van Meerbeek, B. Surface roughness of etched composite resin in light of composite repair. *J. Dent.* **2011**, *39*, 499–505. [CrossRef]

29. Nassoohi, N.; Kazemi, H.; Sadaghiani, M.; Mansouri, M.; Rakhshan, V. Effects of three surface conditioning techniques on repair bond strength of nanohybrid and nanofilled composites. *Dent. Res. J.* **2015**, *12*, 554–561.

30. Lo Giudice, R.; Puleio, F.; Rizzo, D.; Alibrandi, A.; Lo Giudice, G.; Centofanti, A.; Fiorillo, L.; Di Mauro, D.; Nicita, F. Comparative investigation of cutting devices on bone blocks: An SEM morphological analysis. *Appl. Sci.* **2019**, *9*, 351. [CrossRef]

31. Tedesco, M.; Chain, M.C.; Bortoluzzi, E.A.; Garcia, L.F.R.; Alves, A.M.H.; Teixeira, C.S. Comparison of two observational methods, scanning electron and confocal laser scanning microscopies, in the adhesive interface analysis of endodontic sealers to root dentine. *Clin. Oral Invest.* **2018**, *22*, 2353–2361. [CrossRef] [PubMed]

32. Rashid, H. Application of confocal laserscanning microscopy in dentistry. *J. Adv. Microsc. Res.* **2014**, *9*, 245–252. [CrossRef]

33. Cotes, C.; Cardoso, M.; De Melo, R.M.; Valandro, L.F.; Bottino, M.A. Effect of composite surface treatment and aging on the bond strength between a core build-up composite and a luting agent. *J. Appl. Oral Sci.* **2015**, *23*, 71–78. [CrossRef] [PubMed]

34. Ozcan, M.; Cura, C.; Brendeke, J. Effect of aging conditions on the repair bond strength of a microhybrid and a Nanohybrid resin composite. *J. Adhes. Dent.* **2010**, *12*, 451–459.

Article

Assessment of the Different Types of Failure on Anterior Cantilever Resin-Bonded Fixed Dental Prostheses Fabricated with Three Different Materials: An In Vitro Study

Adolfo Di Fiore [1,*], Edoardo Stellini [1], Gianpaolo Savio [2], Stefano Rosso [3], Lorenzo Graiff [1], Stefano Granata [1], Carlo Monaco [4] and Roberto Meneghello [3]

[1] Department of Neurosciences, School of Dentistry, Section of Prosthodontics and Digital Dentistry, University of Padova, 35100 Padova, Italy; edoardo.stellini@unipd.it (E.S.); lorenzo.graiff@unipd.it (L.G.); stefigraf64@yahoo.it (S.G.)

[2] Departments of Civil, Environmental and Architectural Engineering, University of Padova, 35100 Padova, Italy; gianpaolo.savio@unipd.it

[3] Departments of Management and Engineering, University of Padova, 36100 Vicenza, Italy; stefano.rosso.3@phd.unipd.it (S.R.); roberto.meneghello@unipd.it (R.M.)

[4] Department of Biomedical and Neuromotor Sciences (DIBINEM), Division of Prosthodontics and Maxillofacial Rehabilitation, Alma Mater Studiorum— University of Bologna, 40121 Bologna, Italy; carlo.monaco2@unibo.it

* Correspondence: adolfo.difiore@unipd.it

Received: 19 May 2020; Accepted: 15 June 2020; Published: 17 June 2020

Abstract: background: resin-bonded fixed dental prosthesis (RBFDP) represents a highly aesthetic and conservative treatment option to replace a single tooth in a younger patient. The purpose of this in vitro study was to compare the fracture strength and the different types of failure on anterior cantilever RBFDPs fabricated using zirconia (ZR), lithium disilicate (LD), and PMMA-based material with ceramic fillers (PM) by the same standard tessellation language (STL) file. Methods: sixty extracted bovine mandibular incisives were embedded resin block; scanned to design one master model of RBFDP with a cantilevered single-retainer. Twenty cantilevered single-retainer RBFDPs were fabricated using ZR; LD; and PM. Static loading was performed using a universal testing machine. Results: the mean fracture strength for the RBFDPs was: 292.5 Newton (Standard Deviation (SD) 36.6) for ZR; 210 N (SD 37.6) for LD; and 133 N (SD 16.3) for PM. All the failures of RBFDPs in ZR were a fracture of the abutment tooth; instead; the 80% of failures of RBFDPs in LD and PM were a fracture of the connector. Conclusion: within the limitations of this in vitro study, we can conclude that the zirconia RBFDPs presented load resistance higher than the maximum anterior bite force reported in literature (270 N) and failure type analysis showed some trends among the groups

Keywords: zirconia; digital dentistry; lithium disilicate; resin bonded bridge; fracture; adhesive restorations; CAD/CAM; PMMA

---

1. Introduction

Traumatic loss [1], or congenital absence of one anterior maxillary incisor [2] in adolescents, requires immediate treatment with temporary or definitive solutions for aesthetic and functional reasons. Resin-bonded fixed dental prosthesis (RBFDP) represents a highly aesthetic and conservative treatment option to replace a single tooth in a younger patient, before implant treatment becomes available [3] or after orthodontic treatment [4]. In literature, survival rates of RBFDPs were 87.7% in medium-term observation [5]. The main factors of failure include debonding [6], secondary caries on the abutment tooth [5], and fracture of the retainers [7]. These causes are determined by two prosthetic characteristics of RBFDPs: retainer designs [3,8] and properties of the materials [9]. The framework designed with two-retainer has been the most used by clinicians and dental technicians because it was considered with higher fracture resistance than the frameworks with one retainer (cantilever). However, several studies demonstrated that two-retainer RBFDPs have a higher fracture rate and lower survival rate than cantilever RBFDPs [10–15]. Traditionally, the framework of RBFDPs is made of metal alloy, but different metal-free materials with more aesthetics and bond strengths are available. Several authors showed excellent longevity of zirconia [16,17] and lithium ceramic [18] cantilever RBFDPs over 10–20 years with few mechanical complications. However, the mechanical complications were different, according to the material. Debonding rate of 8% and one loss of restoration was revealed by Kern et al. [17] for zirconia cantilever RBFDPs; no debonding was recorded by Sailer et al. [18] for lithium disilicate cantilever RBFDPs. However, the assessment of the mechanical performance of prosthetic materials through clinical trials is difficult because of several conditions and characteristics of the patients.

In the last few years, digital technologies have been introduced in dentistry to improve patient comfort, decrease operative time, and reduce clinical treatment [19–22]. The use of scanners, computer-aided design (CAD) software, and computer-aided manufacturing (CAM) machines have opened many possibilities to replicate and fabricate dental prosthesis in different materials.

Therefore, uniform fabrications of cantilever RBFDPs with different materials by the same standard tessellation language (STL) file allow knowing the real performance of these materials, and the possible adverse comportments on the abutment tooth in extreme conditions.

The purpose of this in vitro study was to compare, using the universal testing machine, the fracture strength and the different types of failure on anterior cantilever RBFDPs fabricated using zirconia, lithium disilicate, and PMMA-based material with ceramic fillers by the same STL file.

2. Materials and Methods

Sixty extracted bovine mandibular incisives were stored in physiological saline solution at a temperature between 5 °C and 10 °C [23]. They were embedded in autopolymerizable methacrylate resin block of dimensions 35 × 50 × 14 mm (ProBase Cold, Ivoclar Vivadent, Bologna, Italy), with the cement–enamel junction above 1 mm and orthogonal the resin base. After horizontal preparation of 1–1.5 mm using a diamond chamfer bur on lingual surface near the cementum–enamel junction of each tooth, air-polishing was performed (S2, EMS) with sodium bicarbonate-based powders (EMS) on 60 teeth for 30 s at distance of about 3 centimeters on total surface of each tooth (Figure 1).

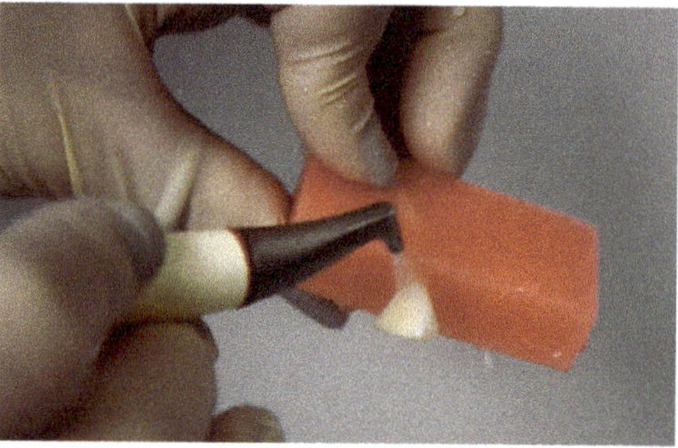

Figure 1. Air-polishing used to clean incisive embedded in resin block.

All teeth were imported in virtual environment by laboratory scanner (Smart Big Open Technology, Rezzato (BS), Italy). Using a Cad software (Exocad DentaCad, Darmstad, Germany), one master model of RBFDP with a cantilevered single-retainer was designed (Figure 2).

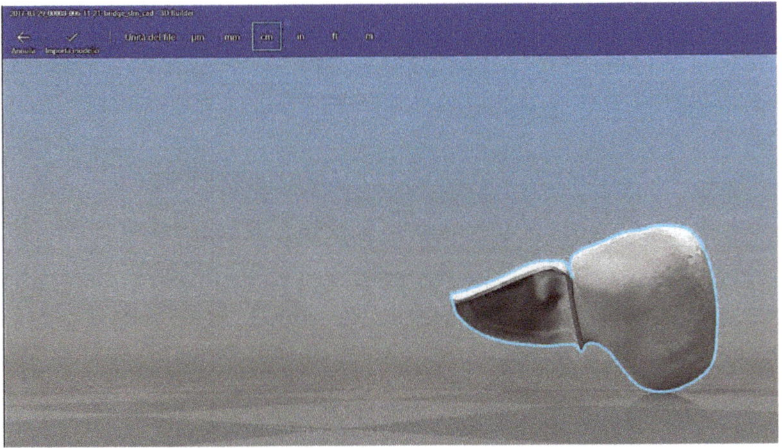

Figure 2. Design of master model of cantilever BFDP.

The RBFDP presented the same parameter setting, shape, and size of the cantilever tooth, independently from the anatomical morphology of the bovine mandibular incisives. The retainer wings were fabricated with a uniform thickness of 0.7 mm. The connectors were designed with a height of 3 mm and a width of 1.5 mm [24,25]. Twenty RBFDPs with a cantilevered single-retainer design for the groups were fabricated with 3 different materials: zirconia (Katana ML, Kuraray, Milano, Italy), lithium disilicate (IPS e.max Press LT A2, Ivoclar Vivadent, Bologna, Italy), and PMMA-based material with ceramic fillers (HIPC, Bredent GmbH, Senden Germany). Each RBFDP has been created by the same CAD design with different procedures based on the type of material. The twenty RBFDPs in HIPC were milled by a computer numerical control machine (Roland DWX-50, Roland DG Corporation, Osaka, Japan). The excessive material was removed with tungsten bur (Komet Dental; H250NEX). All of the samples were polished using a polisher machine (Sirio Dental, Meldola (FC), Italy). Twenty zirconia RBFDPs were milled by a CNC machine (Roland DWX-50, Roland DG Corporation, Osaka, Japan), polished using diamond burs, and sintered (Nabertherm LHT 01/17 D, Nabertherm, Lilienthal, Germany) for 12 h at 1450 °C following the manufacturing instructions. The 20 RBFDPs in lithium disilicate were fabricated with wax lost technique. A wax block (CeraWax, Co.N.Ce.P.T, Busseto (PR), Italy) was used to mill the RBFDPs. A sprue was waxed onto the cantilever tooth of all RBFDPs. They were then embedded in a universal investment material (IPS PressVest Premium investment, Ivoclar Vivadent, Bologna, Italy), with 2 RBFDPs (Figure 3).

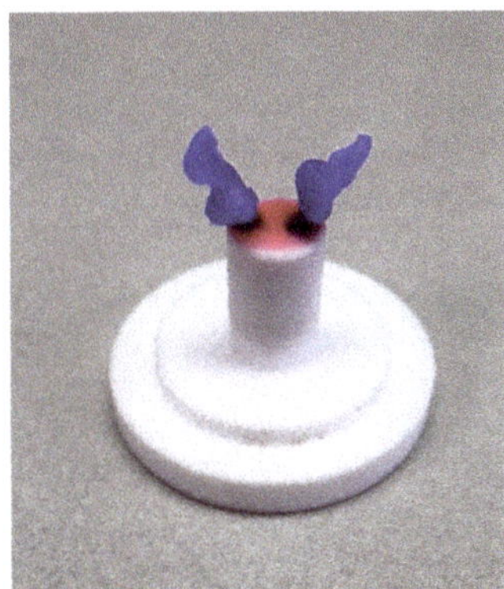

Figure 3. Cantilever RBFDPs milled in wax.

The wax was removed in a heated furnace (Sirio SR 750-In Fire, Sirio Dental S.r.l., Meldola (FC), Italy) and the RBFDPs were pressed with lithium disilicate glass-ceramic (IPS e-max Press LT A2, Ivoclar Vivadent, Bologna, Italy) in the machine-calibrated furnace (Luxor Press-SR 862, Sirio Dental S.r.l., Meldola (FC), Italy) following manufacturer recommendations (Figure 4).

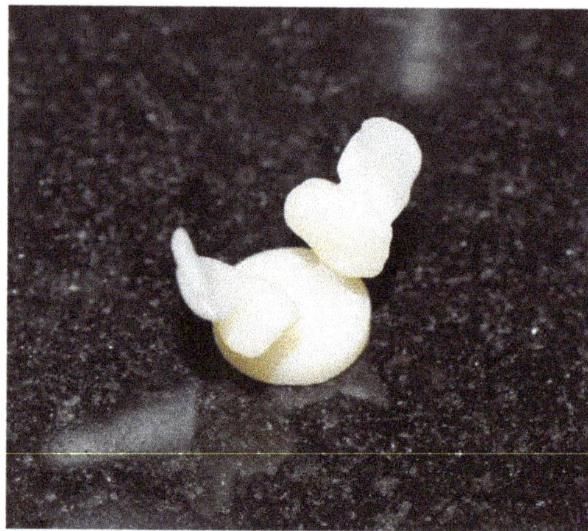

Figure 4. RBFDPs in lithium disilicate fabricated with wax lost technique.

The sprue was removed using a metallic disk and all RBFDPs in lithium disilicate were polish with diamond burs. We obtained 60 comparable RBFDPs with a cantilever design without preparation of the abutment in three different materials. The thickness of the wings of all RBFDPs were measured using a caliper. All of the abutment teeth were treated with the same procedure: 15 s of etching (Scotchbond Universal Etching, 3M, Milano, Italy), 15 s of rinsing with water, 15 s of drying, and the bonding (Scotchbond Universal Adhesive, 3M, Milano, Italy) were applied for 20 s and polymerized for 60 s using a lamp (Valo, Ultradent, South Jordan, UT, USA). All of the RBFDPs were cemented with dual-cured resin cement (Relyx Ultimate, 3M, Milano, Italy) and polymerized for 60 s using led lamp (400–500 nm), but with different procedures. Air abrasion was performed with 50 μm Al_2O_3 particles at a 2.5 bar pressure for 15 s at a distance of 10 mm on the wings of RBFDPs in zirconia. After drying, the wings were cleaned with 96% isopropanol for 3 minutes. The bonding (Scotchbond Universal Adhesive, 3M, Milano italy) was applied on the wings for 20 s and polymerized for 60 s. Air abrasion was performed with 110 μm Al_2O_3 particles at a 2.5 bar pressure for 15 s, at a distance of 10 mm on the wings of RBFDPs in PMMA-based material with ceramic fillers. After drying, the primer (Visiolink, Bredent, Seden, Germany) was applied on the wings and polymerized for 60 s. Instead, the wings of RBFDPs in lithium disilicate were etching with hydrofluoric acid (Ceramic Etching gel, Ivoclar Vivadent, Bologna Italy) for 60 s, 60 s of rinsing, and 30 s of drying. The bonding (Scotchbond Universal Adhesive, 3M, Milano, Italy) was applied for 20 s and polymerized for 60 s. All procedures were performed according to the manufacturer's instructions. The samples were numbered through the engraving on the resin of a code indicating the material. All samples were mounted on a 30° angled support of a universal testing machine (Acumen 3, MTS Systems Corporation, Eden Praire, MN, USA) and a static loading at a crosshead speed of 1.5 mm/min, until failure to the incisal edge of the pontic tooth was performed to simulate real situation (Figure 5). The load was transferred through a 6-mm diameter steatite ball in the middle of the incisal edge of the pontic tooth until failure [25].

Four types of failure were recorded: debonding of RBFDPs, fracture of the connector, debonding of the RBFDPs with fracture of the abutment tooth, and fracture of the abutment tooth with the RBFDP still bonded. All of the methods were synthetized in the following flowchart (Figure 6).

Figure 5. A sample mounted on a 30° angled support of the universal testing machine.

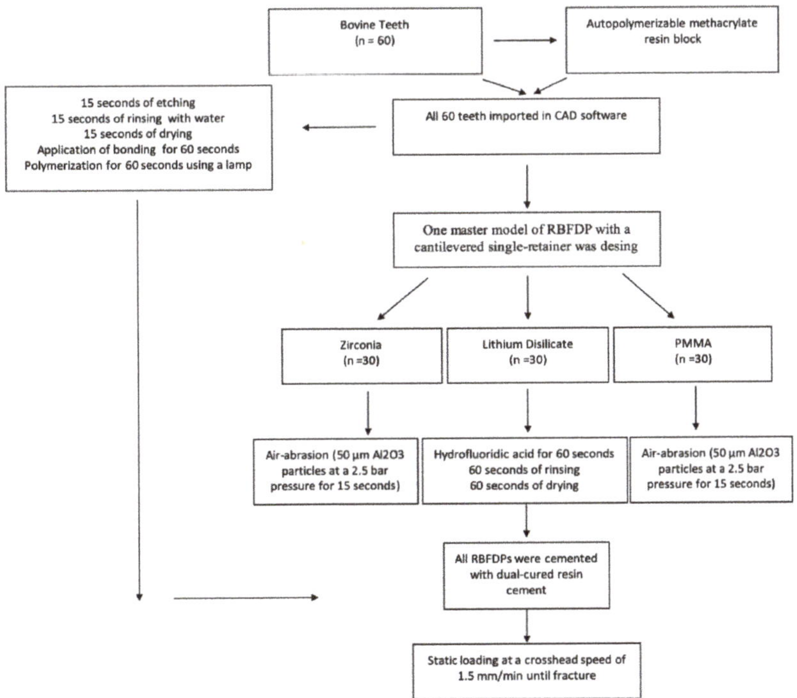

Figure 6. Flowchart of the methods.

The Kruskal–Wallis test with a post hoc analysis using Dunn's test was used to compare the three groups. The level of statistical significance was set as α =0.05 and statistical power of 80%. A statistical software (SPSS v16.0; SPSS Inc., Chicago, IL, USA) was used for the analysis.

3. Results

The mean fracture strength of the RBFDPs in zirconia was 292.5 N (SD 36.6) (Figure 7a), 210 N (SD 37.6) for lithium disilicate (Figure 7b), and 133 N (SD 16.3) for PMMA-based material with ceramic fillers (Figure 7c).

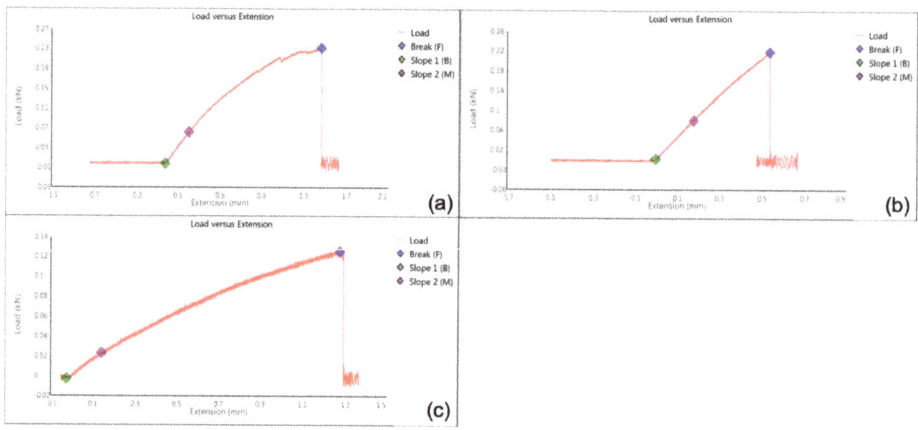

Figure 7. Graphic load-extension of the sample: (**a**) number 3 in zirconia, (**b**) number 5 in lithium disilicate, and (**c**) number 2 in PMMA-based material with ceramic fillers.

In the group of zirconia RBFDPs, all of the failures were fractures of the abutment tooth (Figure 8a); in the group lithium disilicate 80% of failures were fractures of the connector and 20% debonding of the RBFDPs, with fractures of the abutment tooth (Figure 8b). In the group of PMMA-based material with ceramic fillers, 80% of failures were fractures of the connector (Figure 8c) and 20% debonding. Kruskal–Wallis generated a *P*-value < 0.001, identifying statistically significant differences between groups. Dunn's post hoc tests indicated, however, that the difference was statistically significant between the group RBFDPs in zirconia and PMMA-based material with ceramic fillers (P = 0.003), instead no differences were found between RBFDPs in zirconia and lithium disilicate (P = 0.43), and RBFDPs in lithium disilicate and PMMA-based material with ceramic fillers (P = 0.09).

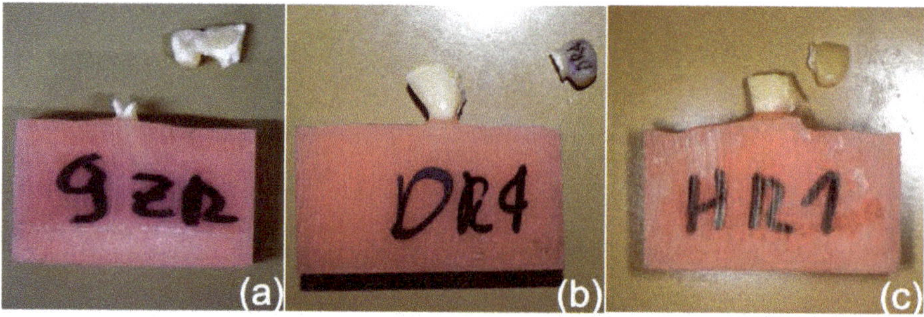

Figure 8. Different types of failure: (**a**) fracture of the abutment tooth in the group RBFDP in zirconia, (**b**) fracture of the connector in the group RBFDP in lithium disilicate, and (**c**) fracture of the connector in the group RBFDP in PMMA-based material with ceramic fillers.

4. Discussion

RBFDP is an aesthetic and conservative treatment option to replace a single tooth in a younger patient; therefore, it is a technique sensitive procedure because it requires proper planning of the clinical case and choice of materials. This in vitro study investigated the different types of failure after static fracture strength tests on RBFDPs fabricated using zirconia, lithium disilicate, and PMMA-based material with ceramic fillers. Zirconia, lithium disilicate, and PMMA-based material with ceramic fillers presented the mean values static fracture strength of 292.5 N (SD 36.6), 210 N (SD 37.6), and 133 N (SD 16.3), respectively. Moreover, different types of failure were collected according to the materials. The 80% of the RBFDPs made in lithium disilicate and PMMA-based material with ceramic fillers presented fracture of the connector; 20% of debonding and all of the RBFDPs in zirconia presented fracture of the abutment tooth.

The results can be explained by the different elastic modulus of materials. The lithium disilicate and PMMA-based material with ceramic fillers presented lower elastic modulus and, consequently, a greater deformation until the fracture of the materials. Another possible explanation of the fracture of the connector was recognizable to the superior bond strength, with respect to the materials. The authors highlight that the lithium disilicate is not indicated to fabricate RBFDPs by manufacturer; however, the results showed a mean fracture strength of 210 N. Therefore, clinical studies are necessary to evaluate the performance of this material used to fabricate RBFDPs. Instead, all the RBFDPs in zirconia presented a fracture of the abutment tooth. The reasons may be attributable to the high elastic modulus of the material or high bond strength of resin luting cement, with respect to the abutment tooth.

Opposite results were observed in clinical trials [16–19] compared to this in vitro study. The reasons could be several; however, the load strength of the RBFDP is different in each patient. Moreover, the pontic elements may not have static and dynamic contacts. In this in vitro study, the uniform fabrication of all the RBFDPs with different materials by the same STL file, and the strength applied through a universal machine until failure, allowed to know the mechanic performance of these prostheses with respect to an in vivo study.

In literature, the same authors in two in vitro studies evaluated the influence of framework design [14] and mode of loading on the fracture strength [26] on cantilever RBFDPs fabricated in pre-sintered aluminum-oxide blocks (In-Ceram alumina blanks). The microscopic examination revealed that 58% of the specimens fractured at the connector only, exactly at the framework-to-veneer interface, 17% fractured at the connector, including veneered parts of the pontic, and 25% fractured at the retainer only. The possible difference of the performance should be explained that, in this research, the authors used bovine teeth. Physiologic occlusal forces for adults in the anterior region were determined to be in a range of 10 to 35 N [27]. Maximal incisive biting forces may vary up to 270 N, primarily depending on facial morphology and age [28]. Compared to the maximal incisive biting forces, only zirconia RBFPDs reached values higher than the physiologic values. Even though the mean fracture strength of lithium disilicate RBFDPs was lower compared to the maximal incisive biting force, it is possible to use the material to fabricate RBFDPs. However, the clinicians should evaluate, with attention, the static and dynamic contacts of the clinical case before use of lithium disilicate RBFDP. Whereas, PMMA-based material with ceramic filler RBFDPs represents a cheap and aesthetic solution for a temporary prosthesis for patients waiting for implant treatments. Therefore, the choice of a correct material, according to therapeutic needs, is crucial for the survival of RBFDPs.

Long-term clinical evidence is needed to evaluate the performance of these materials, especially for use of polymer with ceramic fillers. The drawbacks of this in vitro study were the lack of clinical conditions, as artificial aging, dynamic loading, and of physiologic tooth mobility; however, the uniform fabrications of all the RBFDPs with different materials by the same STL file and the same laboratory conditions allowed to know the real performance of this material, with respect to an in vivo study where the RBFDPs have different shapes and dimensions.

5. Conclusions

Within the limitations of this in vitro study, we can conclude that the zirconia RBFDPs presented load resistance higher than the maximum anterior bite force; furthermore, the lithium disilicate RBFDPs showed a mean fracture strength similar the anterior bite force.

Author Contributions: Conceptualization, A.D.F. and E.S.; methodology, R.M.; validation, S.G. and G.S.; formal analysis, S.G. and S.R.; investigation, G.S. and L.G.; data curation, L.G.; writing—original draft preparation, A.D.F. and E.S.; writing—review and editing, C.M. All authors have read and agreed to the published version of the manuscript.

Funding: This research received no external funding.

Acknowledgments: The authors thank Giorgio Calgaro for materials used for experiments and technical support.

Conflicts of Interest: The authors declare no conflict of interest.

References

1. Borum, M.K.; Andreasen, J.O. Therapeutic and economic implications of traumatic dental injuries in Denmark: An estimate based on 7549 patients treated at a major trauma centre. *Int. J. Paediatr. Dent.* **2001**, *11*, 249–258. [CrossRef]
2. Andrade, D.C.; Loureiro, C.A.; Araujo, V.E.; Riera, R.; Atallah, A.N. Treatment for agenesis of maxillary lateral incisors: A systematic review. *Orthod. Craniofac. Res.* **2013**, *16*, 129–136. [CrossRef] [PubMed]
3. Wei, Y.R.; Wang, X.D.; Zhang, Q.; Li, X.X.; Blatz, M.B.; Jian, Y.T.; Zhao, K. Clinical performance of anterior resin-bonded fixed dental prostheses with different framework designs: A systematic review and meta-analysis. *J. Dent.* **2016**, *47*, 1–7. [CrossRef]
4. Robertsson, S.; Mohlin, B. The congenitally missing upper lateral incisor: A retrospective study of orthodontic space closure versus restorative treatment. *Eur. J. Orthod.* **2000**, *22*, 697–709. [CrossRef]
5. Thoma, D.S.; Sailer, I.; Ioannidis, A.; Zwahlen, M.; Makarov, N.; Pjetursson, B.E. A systematic review of the survival and complication rates of resin-bonded fixed dental prostheses after a mean observation period of at least 5 years. *Clin. Oral. Implant. Res.* **2017**, *28*, 1421–1432. [CrossRef] [PubMed]
6. Botelho, M.G.; Ma, X.; Cheung, G.J.; Law, R.K.; Tai, M.T.; Lam, W.Y. Long-term clinical evaluation of 211 two-unit cantilevered resin-bonded fixed partial dentures. *J. Dent.* **2014**, *42*, 778–784. [CrossRef] [PubMed]
7. Sasse, M.; Eschbach, S.; Kern, M. Randomized clinical trial on single retainer all-ceramic resin-bonded fixed partial dentures: Influence of the bonding system after up to 55 months. *J. Dent.* **2012**, *40*, 783–786. [CrossRef]
8. El-Mowafy, O.; Rubo, M.H. Resin-bonded fixed partial dentures-a literature review with presentation of a novel approach. *Int. J. Prosthodont.* **2000**, *13*, 460–467.
9. Keulemans, F.; Shinya, A.; Lassila, L.V.; Vallittu, P.K.; Kleverlaan, C.J.; Feilzer, A.J.; De Moor, R.J. Three-dimensional finite element analysis of anterior two-unit cantilever resin-bonded fixed dental prostheses. *Sci. World J.* **2015**, *2015*, 864389. [CrossRef]
10. Kern, M.; Sasse, M. Ten-year survival of anterior all-ceramic resin-bonded fixed dental prostheses. *J. Adhes. Dent.* **2011**, *13*, 407–410.
11. Kern, M. Clinical long-term survival of two-retainer and single-retainer all-ceramic resin-bonded fixed partial dentures. *Quintessence Int.* **2005**, *36*, 141–147. [PubMed]
12. Botelho, M.G.; Leung, K.C.; Ng, H.; Chan, K. A retrospective clinical evaluation of two-unit cantilevered resin-bonded fixed partial dentures. *J. Am. Dent. Assoc.* **2006**, *137*, 783–788. [CrossRef] [PubMed]
13. Botelho, M.G.; Chan, A.W.; Leung, N.C.; Lam, W.Y. Long-term evaluation of cantilevered versus fixed-fixed resin-bonded fixed partial dentures for missing maxillary incisors. *J. Dent.* **2016**, *45*, 59–66. [CrossRef] [PubMed]
14. Mourshed, B.; Samran, A.; Alfagih, A.; Samran, A.; Abdulrab, S.; Kern, M. Anterior cantilever resin-bonded fixed dental prostheses: A review of the literature. *J. Prosthodont.* **2018**, *27*, 266–275. [CrossRef]
15. Kern, M. Fifteen-year survival of anterior all-ceramic cantilever resin-bonded fixed dental prostheses. *J. Dent.* **2017**, *56*, 133–135. [CrossRef]
16. Sasse, M.; Kern, M. Survival of anterior cantilevered all-ceramic resin-bonded fixed dental prostheses made from zirconia ceramic. *J. Dent.* **2014**, *42*, 660–663. [CrossRef]

17. Kern, M.; Passia, N.; Sasse, M.; Yazigi, C. Ten-year outcome of zirconia ceramic cantilever resin-bonded fixed dental prostheses and the influence of the reasons for missing incisors. *J. Dent.* **2017**, *65*, 51–55. [CrossRef]

18. Sailer, I.; Bonani, T.; Brodbeck, U.; Hämmerle, C.H. Retrospective clinical study of single-retainer cantilever anterior and posterior glass-ceramic resin-bonded fixed dental prostheses at a mean follow-up of 6 years. *Int. J. Prosthodont.* **2013**, *26*, 443–450. [CrossRef]

19. Di Fiore, A.; Meneghello, R.; Graiff, L.; Savio, G.; Vigolo, P.; Monaco, C.; Stellini, E. Full arch digital scanning systems performances for implant-supported fixed dental prostheses: A comparative study of 8 intraoral scanners. *J. Prosthodont. Res.* **2019**, *63*, 396–403. [CrossRef]

20. Di Fiore, A.; Vigolo, P.; Graiff, L.; Stellini, E. Digital vs Conventional Workflow for Screw-Retained Single-Implant Crowns: A Comparison of Key Considerations. *Int. J. Prosthodont.* **2018**, *31*, 577–579. [CrossRef]

21. Granata, S.; Giberti, L.; Vigolo, P.; Stellini, E.; Di Fiore, A. Incorporating a facial scanner into the digital workflow: A dental technique. *J. Prosthet. Dent.* **2019**, *19*, 30356–30357. [CrossRef] [PubMed]

22. Malaguti, G.; Rossi, R.; Marziali, B.; Esposito, A.; Bruno, G.; Dariol, C.; Di Fiore, A. In vitro evaluation of prosthodontic impression on natural dentition: A comparison between traditional and digital techniques. *Oral Implantol.* **2017**, *9*, 21–27. [CrossRef] [PubMed]

23. Sirisha, K.; Rambabu, T.; Ravishankar, Y.; Ravikumar, P. Validity of bond strength tests: A critical review-Part II. *J. Conserv. Dent.* **2014**, *17*, 420–426. [CrossRef]

24. Koutayas, S.O.; Kern, M.; Ferraresso, F.; Strub, J.R. Influence of framework design on fracture strength of mandibular anterior all-ceramic resin-bonded fixed partial dentures. *Int. J. Prosthodont.* **2002**, *15*, 223–229.

25. Yokoyama, D.; Shinya, A.; Lassila, L.V.; Gomi, H.; Nakasone, Y.; Vallittu, P.K.; Shinya, A. Framework design of an anterior fiber-reinforced hybrid composite fixed partial denture: A 3D finite element study. *Int. J. Prosthodont.* **2009**, *22*, 405–412. [PubMed]

26. Koutayas, S.O.; Kern, M.; Ferraresso, F.; Strub, J.R. Influence of design and mode of loading on the fracture strength of all-ceramic resin-bonded fixed partial dentures: An in vitro study in a dual-axis chewing simulator. *J. Prosthet. Dent.* **2000**, *84*, 112. [CrossRef]

27. De Boever, J.A.; McCall, W.D., Jr.; Holden, S.; Ash, M.M., Jr. Functional occlusal forces: An investigation by telemetry. *J. Prosthet. Dent.* **1978**, *40*, 326–333. [CrossRef]

28. Kiliaridis, S.; Kjellberg, H.; Wenneberg, B.; Engstrom, C. The relationship between maximal force, bite force endurance, and facial morphology during growth. A cross-sectional study. *Acta. Odontol. Scand.* **1993**, *51*, 323–331. [CrossRef]

MDPI
St. Alban-Anlage 66
4052 Basel
Switzerland
Tel. +41 61 683 77 34
Fax +41 61 302 89 18
www.mdpi.com

Applied Sciences Editorial Office
E-mail: applsci@mdpi.com
www.mdpi.com/journal/applsci